IMMUNOLOGICAL
COMPUTATION

OTHER AUERBACH PUBLICATIONS

Chaos Applications in Telecommunications
Peter Stavroulakis
ISBN: 0-8493-3832-8

Contemporary Coding Techniques and Applications for Mobile Communications
Onur Osman and Osman Nuri Ucan
ISBN: 1-4200-5461-9

Design Science Research Methods and Patterns: Innovating Information and Communication Technology
Vijay K. Vaishnavi and William Kuechler Jr.
ISBN: 1-4200-5932-7

Embedded Linux System Design and Development
P. Raghavan, Amol Lad and Sriram Neelakandan
ISBN: 0-8493-4058-6

Enhancing Computer Security with Smart Technology
V. Rao Vemuri
ISBN: 0-8493-3045-9

Graph Theory and Interconnection Networks
Lih-Hsing Hsu; Cheng-Kuan Lin
ISBN: 1-4200-4481-8

Modeling Software with Finite State Machines: A Practical Approach
Ferdinand Wagner, Ruedi Schmuki, Thomas Wagner and Peter Wolstenholme
ISBN: 0-8493-8086-3

MEMS and Nanotechnology-Based Sensors and Devices for Communications, Medical and Aerospace Applications
A. R. Jha
ISBN: 0-8493-8069-3

Neural Networks for Applied Sciences and Engineering: From Fundamentals to Complex Pattern Recognition
Sandhya Samarasinghe
ISBN: 0-8493-3375-X

Patterns for Performance and Operability: Building and Testing Enterprise Software
Chris Ford, Ido Gileadi, Sanjiv Purba and Mike Moerman
ISBN: 1-4200-5334-5

Software Engineering Foundations: A Software Science Perspective
Yingxu Wang
ISBN: 0-8493-1931-5

Systemic Yoyos: Some Impacts of the Second Dimension
Yi Lin
ISBN: 1-4200-8820-3

AUERBACH PUBLICATIONS
www.auerbach-publications.com
To Order Call: 1-800-272-7737 • Fax: 1-800-374-3401
E-mail: orders@crcpress.com

IMMUNOLOGICAL COMPUTATION

Theory and Applications

Dipankar Dasgupta
Luis Fernando Niño

CRC Press
Taylor & Francis Group
Boca Raton London New York

CRC Press is an imprint of the
Taylor & Francis Group, an **informa** business
AN AUERBACH BOOK

CRC Press
Taylor & Francis Group
6000 Broken Sound Parkway NW, Suite 300
Boca Raton, FL 33487-2742

First issued in paperback 2019

ISBN-13: 978-1-4200-6545-9 (hbk)
ISBN-13: 978-0-367-38690-0 (pbk)

Library of Congress Cataloging-in-Publication Data

Dasgupta, D. (Dipankar), 1958-
    Immunological computation : theory and applications / Dipankar Dasgupta and Luis Fernando Niño.
        p. cm.
    Includes bibliographical references and index.
    ISBN 978-1-4200-6545-9 (alk. paper)
    1. Immunocomputers. 2. Artificial intelligence. 3. Immune system--Computer simulation. I. Niño, Luis Fernando. II. Title.

QA76.875.D37 2009
006.3--dc22   2008013322

Visit the Taylor & Francis Web site at
http://www.taylorandfrancis.com

and the Auerbach Web site at
http://www.crcpress.com

To our parents

Contents

Preface

Immunological computation (also called artificial immune systems [AIS]) is a field of study devoted to the development of computational models based on the principles of the biological immune system (BIS). It is an emerging area that explores and employs different immunological mechanisms to solve computational problems.

The BIS is a complex, adaptive, highly distributive learning system with several mechanisms for defense against pathogenic organisms. It employs several alternative and complementary mechanisms for defense against foreign pathogens. The immune system learns, through adaptation, to distinguish between dangerous foreign antigens and the body's own cells or molecules. Clearly, nature has been very effective in creating organisms that are capable of protecting themselves against a wide variety of pathogens such as bacteria, fungi, and parasites.

The powerful information-processing capabilities of the immune system, such as feature extraction, pattern recognition, learning, memory, and its distributive nature provide rich metaphors for its artificial counterpart. From 1985 to 2007 there has been an increased research interest in immunity-based techniques and their applications. Some of these models, however, are intended to describe the processes that occur in the BIS to have a better understanding of the dynamic behavior of immunological processes and simulate BIS's dynamic behavior in the presence of antigens/pathogens. In contrast, immune-inspired models have been developed in an attempt to solve complex real-world problems such as anomaly detection, pattern recognition, data analysis (clustering), function optimization, and computer security.

This book is devoted to discussing different immunological mechanisms and their relation to information processing and problem solving. This is the first book that can be used as a textbook in the area of immunological computation; it presents a compendium of up-to-date work related to immunity-based techniques. Each chapter provides a summary, review questions, and exercises for students to practice; chapters also include some issues to research further. This book is also suitable as a reference text for graduate study in computational intelligence, bioinspired computing, AIS, and other related areas. *Immunological Computation: Theory and Applications* will be of professional interest to scientists, academics, and practitioners.

This book consists of seven chapters. Chapter 1 summarizes the fundamental concepts of immunology necessary to understand the computational models based on immunology presented in subsequent chapters. Particularly, some components of the BIS such as B cells, T cells, and other lymphocytes are described. Other concepts such as antigen, immune response, and an outline of the global process by which antigen recognition is achieved are also summarized.

Chapter 2 presents some theoretical models of immune processes. Specifically, such theories attempt to explain adaptive immune response. For instance, the immune system's ability to "remember" its encounters with antigens to achieve a faster response when the same antigen is confronted at a later time is being studied. Thus, immune networks and danger theory are discussed. In addition, the main computational aspects of the immune system are highlighted at the end of the chapter.

Chapter 3 presents general abstractions of some immune elements and processes usually used in most computational models. Accordingly, Chapter 3 reviews standard procedures, representations, and matching rules that are used in all immunological computation models.

Chapter 4 covers the details of one of the earliest and most well-known immune algorithms, which is based on the negative selection (NS) process that occurs in the thymus. The chapter presents change/anomaly detection techniques inspired by the T cell censoring and maturation process in the BIS. Thus, first, the main process of NS that results in self–nonself discrimination is described. Then, the important features of the artificial NS are presented, followed by different variations of NS algorithms.

Chapter 5 concentrates on immune networks. At the beginning of this chapter, some immune models algorithms based on clonal selection, and which are very closely related to immune networks, are specified. In the rest of the chapter, the most important continuous and discrete immune network models are detailed. In addition, at the end of this chapter, a generic immune network model is described.

In Chapter 6, some promising immune models, which have been recently proposed, are briefly discussed. Particularly, such models are based on danger theory, cytokine network models, and MHC-based models.

Finally, Chapter 7 highlights how AIS contribute to cross-linking solutions to different real-world problems. It briefly describes a wide variety of applications, which include computer security, fault detection and diagnosis, anomaly detection, robotics, and data mining among others.

At the end of this book, an indexed bibliography of up-to-date publications, events, and researchers in immunological computation and related fields is included.

Acknowledgments

Thanks to all our friends, students, and colleagues for their help and constructive criticisms. Special thanks to our students (Aishwarya Kaushal, Senhua Yu, and Sowmya Sree Veeravatnam, Sudip Saha) at the Artificial Immune Systems Research Laboratory, University of Memphis and the Intelligent Systems Research Laboratory, National University in Bogota.

Dr. Niño thanks his students Oscar Alonso and David Becerra from the Computer Science Department, National University of Colombia for their valuable help and suggested corrections to the manuscript. Dr. Dasgupta would like to give special thanks to his wife (Geeta) and children (Sonali and Sukanya) for bearing with him, giving moral support and encouragement in finishing this book.

Dipankar Dasgupta
Luis Fernando Niño

Authors

Dipankar Dasgupta is a professor of computer science at the University of Memphis. His research interests are in the broad area of scientific computing and tracking real-world problems through interdisciplinary cooperation. His areas of special interests include AIS, genetic algorithms, neural networks, and multiagent systems and their applications.

He obtained his BS in electrical engineering (1981) from the Assam Engineering College, Guwahati, and MSc in computer engineering (1986) from the Indian Institute of Technology, Kharagpur, from India and his PhD in computer science (1994) from the University of Strathclyde, Glasgow. He has published more than 130 research papers in books, journals, and international conferences. He has edited two books: one is on genetic algorithms and the other entitled *Artificial Immune Systems and Their Applications* published by Springer, 1999. The book on AIS is the first book in the field and is widely used as a reference book.

Dr. Dasgupta is a senior member of IEEE and ACM and regularly serves as a panelist, keynote speaker, and program committee member (five to six per year) at many international conferences. He first started (in 1997) organizing special tracks and workshops on AIS and has regularly offered tutorials on the topics at international conferences since then. Dr. Dasgupta edited a special issue (on AIS) of *IEEE Evolutionary Computation Journal* (Volume 6, Number 3, June 2002). Since 1998, he has been an associate editor of the journal *IEEE Transactions on Evolutionary Computation*. He also was nominated as the chair of IEEE Task Force on AIS. His research lab regularly updates AIS bibliography and publishes it on the web (available at http://ais.cs.memphis.edu.).

Luis Fernando Niño is an associate professor in the Computer and Industrial Engineering Department, National University of Colombia, Bogota. His main areas of interest include bioinspired computing, specifically evolutionary computation, neural networks, AIS, and swarm intelligence. He has published more than

20 research papers. He obtained his BS in computer engineering (1990) and his MSc in mathematics (1995) from the National University of Colombia and his MSc in computer science (1999) and his PhD in computer science (2000) from the University of Memphis. Dr. Niño completed his dissertation work under the supervision of Professor Dipankar Dasgupta and has visited the latter as a postdoctoral researcher several times at the University of Memphis since he graduated.

Chapter 1

Immunology Basics

In medicine, historically, the term "immunity" refers to the condition in which an organism can resist diseases, more specifically infectious diseases. However, a broader definition of immunity is the reaction to foreign substances (pathogens), which includes primary and secondary immune responses.

Mammals have developed a robust defense system called the *immune system* to deal with foreign and potentially dangerous pathogens. The immune system consists of a set of organs, cells, and molecules; and their coordinated response in the presence of a pathogen is known as the *immune response*. In a broader sense, the physiological function of the immune system is to defend an organism against all kinds of harmful substances such as fungi, bacteria, parasites, viruses, and other protozoa. However, noninfectious external substances can also generate immune responses (Abbas and Lichtman, 2005).

In general, antigens are capable of inducing an immune response as they are assumed to be harmful nonself invaders in the body. The ability of an antigen to induce an immune response probably depends on four main factors: foreignness, molecular size, chemical composition and heterogeneity, and susceptibility to antigen processing and antigen presentation.

The biological immune system (BIS) has the ability to detect foreign substances, and to respond adequately. It is inherently distributed and fault-tolerant, and exhibits a complex behavior while interacting with all its constituents. One of the main capabilities of the immune system is to distinguish own body cells from foreign substances, which is called *self/nonself discrimination*. In general, the BIS is capable of recognizing the dangerous elements and deciding an appropriate response while tolerating self-molecules and ignoring many harmless substances.

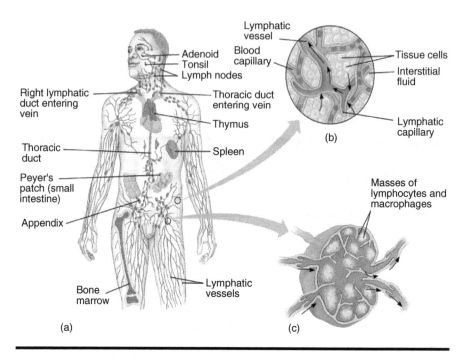

Figure 1.1 Functional Components of the immune system.

1.1 Functional Elements of the Immune System

The immune system is a collection of organs, cells, and molecules responsible for dealing with potentially harmful invaders; it also has other functionalities in the body.

1.1.1 Organs

The organs, which constitute the immune system, can be classified into central lymphoid organs and peripheral lymphoid organs. The purpose of central lymphoid organs is to generate and assist mature immune cells (lymphocytes). Such organs include the bone marrow and the thymus. However, peripheral lymphoid organs facilitate the interaction between lymphocytes and antigens, as the antigen concentration increases in these organs. Peripheral lymphoid organs include lymph nodes, the spleen, and mucosal and submucosal tissues of the alimentary and respiratory tracts. Figure 1.1 shows the components of the biological immune system.

1.1.1.1 Bone Marrow

In an abstract sense, naive immune cells are initially generated in the bone marrow and are derived through a process called hematopoiesis. During hematopoiesis, bone marrow–derived stem cells divide into either mature immune cells (to perform immunological function) or precursors of cells that migrate out of the bone marrow to continue their maturation process elsewhere (thymus or germinal center (GC)). In addition to red blood cells and platelets, the bone marrow produces B cells, natural killer cells, granulocytes, and immature thymocytes.

1.1.1.2 Thymus

In simple terms, the function of the thymus is to produce mature T cells. Some immature immune cells (thymocytes), also known as prothymocytes, leave the bone marrow and migrate into the thymus. Through a maturation process, sometimes referred to as "thymic education," T cells that are beneficial to the immune system are kept, whereas those T cells that might cause a detrimental autoimmune response are eliminated; mature T cells are then released into the bloodstream for performing immunological functions.

1.1.1.3 Spleen

The spleen is an organ, which is made up of B cells, T cells, macrophages, dendritic cells, natural killer cells, and red blood cells. In addition to capturing foreign substances (pathogens) from the blood that passes through the spleen, migratory macrophages and dendritic cells bring antigens to the spleen through the bloodstream. An immune response is initiated when macrophages or dendritic cells present the antigen to the appropriate B or T cells. This organ can be thought of as an immunological "conference center." In the spleen, B cells become activated and produce large amounts of antibodies in one of its factories, called the general center. Additionally, old red blood cells are destroyed in the spleen.

1.1.1.4 Lymph Node

The function of lymph nodes is to act as an immunologic filter for the fluid known as lymph. Lymph nodes are found throughout the body and they are mostly composed of T cells, B cells, dendritic cells, and macrophages. Such nodes drain fluid from most of the body tissues. Antigens are filtered out of the lymph (a fluid that contains white blood cells) in the lymph nodes before returning the lymph to circulation throughout the lymphatic system. Similar to the spleen, macrophages and dendritic cells that capture antigens present these to T and B cells, thus initiating an immune response.

1.1.2 Immune Cells and Molecules

The immune system is composed of a variety of cells and molecules, which interact among themselves to achieve appropriate immunological responses (biological defense).

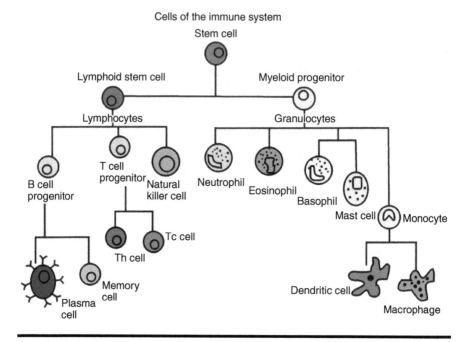

Figure 1.2   Immune cells that contribute to immune response.

Some of those cells that take part in immune response are shown in a hierarchical diagram in Figure 1.2. The most relevant ones are described in some detail in the following sections.

1.1.2.1   Lymphocytes, T Lymphocytes, and B Lymphocytes

White blood cells, also called lymphocytes, are very important constituents of the immune system. These cells are produced in the bone marrow, circulate in the blood and lymph system, and reside in various lymphoid organs to perform immunological functions. The primary lymphoid organs provide sites where lymphocytes mature and become antigenically committed. B and T cells constitute the major population of lymphocytes.

T cells are specialized cells of the immune system, which are matured in the thymus. The thymus produces five subpopulations of T cells as follows:

■ Delayed hypersensitivity T cells, which are a type of T cells that produce cytokines that direct the cellular-mediated immune response and phagocytosis.
■ Helper T cells, which help the B cells to perform antigen recognition by releasing cytokines.
■ Cytotoxic T cells, which kill infected self-cells and tumor cells. They also kill foreign cells.

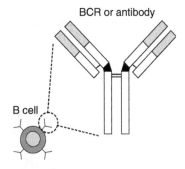

Figure 1.3   Illustration of B cell receptors—B cells have immunoglobulin receptors on their surface, which bind to antigens.

- Memory T cells, which form a pool that will remember earlier immune responses.
- Suppressor T cells, which inhibit an immune response by suppressing helper T cells. As a result, specific antibodies will not be produced. These serve to suppress false alarms.

As mentioned earlier, B cells are another important class of immune cells, which can recognize particular antigens. There are billions of these cells circulating the body, constituting an effective and distributed anomaly detection and response system (Clancy, 1998; Kuby et al., 2000; Sompayrac, 2003). B cells are specialized white blood cells produced in the bone marrow and are responsible for producing and secreting Y-shaped antibodies, which bind to antigens (see Figure 1.3). Each B cell secretes multiple copies of one kind of antibody for antigen match. Activated B cells become memory cells or plasma cells; the latter actively secret antibodies.

1.1.2.2   Antibodies

Antibodies (Abs) are a particular kind of molecules, called immunoglobulins found in the blood and produced by mature B cells, also known as plasma cells.

An antibody contains four polypeptide chains: two identical *light* chains and two identical *heavy* chains. Each chain comprises a variable region (*V*) and a constant region (*C*) (see Figure 1.4). Both *V* regions combine to form two antigen-binding sites, also known as antigen-binding regions (ABR).

1.1.2.3   Cytokines, Lymphokines, and Interleukins

Cytokines are a group of proteins and peptides that get secreted by some immune cells to influence the behavior of other cells. These are chemical messengers allowing intercellular communication by binding to the membrane of a target cell.

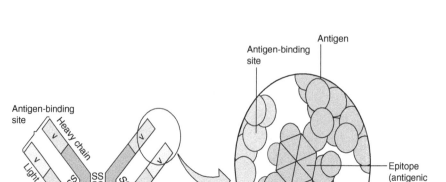

Figure 1.4   Details of an antibody molecule—surface Ig contains four polypeptide chains: two identical *light* chains and two identical *heavy* chains. Each chain comprises a variable region (*V*) and a constant region (*C*). Both *V* regions combine to form two antigen-binding sites. (a) Antibody molecule, (b) enlarged antigen-binding site.

Cytokines are mainly secreted by lymphocytes; however, they can also be produced by other immune and nonimmune cells, for example, those cells that secrete cytokines when they get damaged (Abbas and Lichtman, 2004; Baumann and Gauldie, 1994). Cytokines secreted by lymphocytes are called *lymphokines* and they have a strong influence over immune processes. Lymphokines produced by a lymphocyte to communicate with other lymphocytes are called interleukins.

1.1.2.4   Peptides, Major Histocompatibility Complex, and Antigen Presenting Cells

The term peptide refers to a short chain of amino acids, usually obtained by the fragmentation of an antigen, and presented to other cells of the immune system by antigen presenting cells (APC).

Antigen presentation refers to processing a suspicious foreign particle. Such a particle is broken up into peptides, and then such peptides are held on the surface of APC, where T cells can recognize them. Several types of cells may serve as APC, including macrophages, dendritic cells, and B cells. For instance, macrophages act as phagocytes which engulf foreign antigens, and then antigens become internalized, processed, and expressed on the macrophages' surface.

In contrast, major histocompatibility complex (MHC) proteins act as "sign posts" that display peptides on the surface of a host cell. All MHC molecules receive polypeptides from inside the cells they are a part of, and display them on the cell's exterior surface for recognition by T cells.

MHC proteins can be classified into three classes: MHC class I molecules are found on almost every nucleated cell of the body. Class I molecules present "endogenous" antigens to cytotoxic T cells. Endogenous antigens may be fragments of viral proteins or tumor proteins. Presentation of such antigens would indicate internal cellular alterations, which if not contained could spread throughout the body.

MHC class II molecules are found only on a few specialized cell types, including macrophages, dendritic cells, activated T and B cells. Moreover, class II molecules present "exogenous" antigens to helper T cells. Exogenous antigens might be fragments of bacterial cells or viruses that are engulfed and processed by, for example, a macrophage and then presented to helper T cells.

MHC class III region encodes for other immune components such as complement components (e.g., C2, C4, or factor B) and some that encode cytokines (e.g., TNF-α).

1.1.2.5 Macrophages and Dendritic Cells

Macrophages are specialized cells, which engulf large particles such as bacteria, yeast, and dying cells by a process called phagocytosis. When a macrophage ingests a pathogen, the pathogen becomes trapped in a food vacuole, which then fuses with a lysosome. Enzymes and toxic oxygen compounds digest the invader within the lysosome.

Dendritic cells are immune cells that form part of the mammal immune system. These cells are present in small amounts in those tissues that are in contact with the external environment such as the skin (where they are often called Langerhans cells) and the inner covering of nose, lungs, stomach, and intestines. In their immature state, they can also be found in blood. Once activated, they migrate to the lymphoid tissues, where they interact with T- and B cells to initiate and drive an immune response.

1.1.3 The Complement System

The complement system is a part of humoral immunity; when an infection occurs, this system complements the antibacterial activity of antibody. It consists of a set of plasma proteins; if one gets activated, it triggers a sequence of reactions on a pathogen's surface that helps to destroy the pathogen and eliminate the infection. The three main consequences of the complement system are the recruitment of inflammatory immune cells (phagocytes), opsonization of pathogens by antibodies, and killing of pathogens by creating pores in the bacterial membrane leading to their death. So the complement activation helps to amplify the effects of the classical pathway.

Nonspecific defense mechanisms		Specific defense mechanisms
First line of defense: Anatomic barrier	Second line of defense: Innate immunity	Third line of defense: Adaptive immunity
Skin Mucous membranes Secretions of skin and mucous membranes	Phagocytic white blood cells Antimicrobial proteins Inflammatory response	Lymphocytes Antibodies

Figure 1.5 Biological defense mechanisms—nonspecific and specific defense mechanisms.

1.2 Layers of the Immune System

The immune system can be envisioned as a multilayer system, each layer consisting of different types of defense mechanisms (Kuby et al., 2000; Pathak and Palan, 2005). The three main layers include the anatomic barrier, the innate immunity, and the adaptive immunity.

Biological defense mechanisms may be classified into two categories: nonspecific and specific defense mechanisms (see Figure 1.5). Nonspecific defense mechanisms produce the same type of response independent of pathogen that enters the body. In contrast, specific defense mechanisms are based on recognizing particular pathogens. Each one of these defense mechanisms is explained in the following sections.

1.2.1 Anatomic Barrier

The first layer of the biological defense is the anatomic barrier, composed of the skin and the surface of mucous membranes. Intact skin prevents the eruption of most pathogens and also inhibits most bacterial growth because of its low pH. In contrast, many pathogens enter the body by binding or penetrating through the mucous membranes; thus, the role of these membranes is to provide a number of nonspecific mechanisms that help prevent such invasions. For example, saliva, tears, and some mucous secretions, which contain antibacterial and antiviral substances (Kuby et al., 2000; Sompayrac, 2003), wash away potential invaders.

1.2.2 Innate Immunity

Innate immunity (Woods, 1991) refers to all defense mechanisms against foreign pathogens that individuals are born with. Innate immunity is mainly composed of the following mechanisms:

■ *Phagocytic barriers.* Some specialized cells (like macrophages, neutrophils, and natural killer cells) are able to ingest foreign substances, including whole pathogenic microorganisms. This ingestion has two purposes: to kill the antigen and to present fragments of the invader's proteins to other immune cells and molecules.

■ *Inflammatory response.* Activated macrophages produce cytokines (hormone-like protein messengers), which induce the inflammatory response characterized by vasodilation and rise in capillary permeability. These changes allow a large number of circulating immune cells to be recruited to the site where an infection occurs.

1.2.3 Adaptive Immunity

Adaptive immunity (Kuby et al., 2000; Stanley, 2002), also called acquired or specific immunity, represents the part of the immune mechanism that is able to specifically recognize and selectively eliminate foreign microorganisms and molecules. Adaptive immunity produces two types of responses in the presence of pathogens: *humoral* immunity and *cellular* immunity. The humoral immunity is based on the synthesis of antibodies by B cells; however, in cellular immunity, T cells cause the destruction of microorganisms that carry invading antigens and those self-cells that have been infected.

■ *Humoral immunity.* Humoral immunity is mediated by antibodies contained in body fluids (known as humors). The humoral branch of the immune system involves B cell/antigen interaction, and the subsequent proliferation and differentiation of B cells into antibody-secreting plasma cells. Antibodies function as effectors of the humoral response by binding to antigens and facilitating their elimination.

■ *Cellular immunity.* Cellular immunity is cell-mediated; thus, effector T cells, generated in response to an antigen, are responsible for cell-mediated immunity. Cytotoxic T lymphocytes (CTLs) participate in cell-mediated immune reactions by killing altered self-cells; they play an important role in killing virus-infected- and tumor cells. Cytokines secreted by T_{DH} cells can mediate cellular immunity, and activate various phagocytic cells, enabling them to kill microorganisms more effectively. This type of cell-mediated immune response is especially important in host defense against intracellular bacteria and protozoa (Abbas and Lichtman, 2004; Todd and Spickett, 2005).

1.3 Immune System Dynamics

The mechanisms that define the immune system's dynamic behavior are explained in this section.

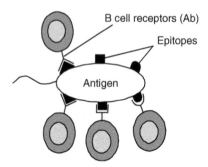

Figure 1.6 Antigen recognition by B cells. Immune recognition is based on the complementarity between the binding region of a receptor and an antigen epitope.

1.3.1 Immune Recognition: Matching and Binding

Several immunological processes require an element (cell or molecule) of the immune system to recognize the presence of another element. T cell recognition is based on the complementarity between the binding region of the cell molecule and the receptor. For instance, antigens are detected when a molecular bond is established between the antigen and receptors on the surface of B cells (see Figure 1.6). Because of the large size and complexity of most antigens, only parts of the antigen, discrete sites called *epitopes*, get bound to B cell receptors. Multiple receptors bind to an antigen with varying affinity, that is, the more complementary the structures of the epitope and the B cell receptor are, the more likely for a stronger bond to occur.

Accordingly, binding or detection in the immune system is approximate to stimulate a primary response. This is probably because it is too difficult to evolve receptor structures that are exact complementary to epitopes (antigen) never encountered before. If precise binding were required, the chances of a random lymphocyte binding to a random epitope would be small. An important consequence of approximate binding is that a single lymphocyte can detect a subset of epitopes, which means that fewer lymphocytes are needed to provide protection against a variety of possible pathogens. This feature makes the immune system efficient in terms of time and memory.

A lymphocyte has approximately 10^5 receptors on its surface; because all of these receptors have the same structure (i.e., a lymphocyte is monoclonal), a single lymphocyte can only bind to structurally related epitopes. These structurally related epitopes define the similarity subset that a lymphocyte can detect. The number of receptors that bind to pathogens determines the affinity of the lymphocyte toward a given pathogen. If a bond is very likely to occur, then many receptors may bind to pathogen epitopes, resulting in a high affinity for that pathogen. However, if a bond is unlikely to occur, then few receptors might bind to epitopes, and the lymphocyte will have low affinity for that pathogen. If the lymphocyte's affinity for the pathogen exceeds a certain *affinity threshold*, it sends signals to other immune cells, which results in immune response. As the affinity threshold increases, the number of epitope types that can activate the lymphocyte decreases, that is, the similarity subset becomes smaller.

1.3.2 Response to Antigens

The response to the presence of antigens is composed of two interlinked mechanisms: innate immunity and adaptive immunity.

The first one, innate immunity, is achieved by some specialized cells (like macrophages, neutrophils, and natural killer cells) that are able to ingest and kill foreign substances, including whole pathogenic microorganisms. Activated macrophages produce cytokines, which induce the inflammatory response, characterized by vasodilation and rise in capillary permeability. These changes allow a large number of circulating immune cells to be recruited to the infected site. Innate immunity provides a fast response against antigens in contrast to adaptive immunity.

When pathogenic microorganisms (e.g., pathogen, virus, and parasites) invade an organism, T_{DH} cells can recognize the infection and produce cytotoxic factor (CF). This tells the macrophage to track pathogens at that site. After finding pathogens, T_{DH} cells produce migration inhibitory factor (MIF) for macrophages to refrain them from leaving the reaction site.

In contrast, adaptive (specific) immunity is divided into humoral immunity and cellular immunity. Humoral immunity amplifies the innate immune response by producing antibodies. Then, microorganisms are coated with antibodies or complement products so that they can easily be recognized (opsonization). Opsonization means "preparation for eating" and accordingly, extracellular material is then ingested by macrophages. Adaptive immunity requires the development of antibodies, which are specific to each antigen.

Humoral immune response has the following phases: a macrophage ingests an antigen and becomes an APC. This APC stimulates helper T cells, which then secrete lymphokines. Subsequently, when a B cell recognizes an antigen with the presence of lymphokines secreted by helper T cells, it differentiates into a plasma cell or a memory cell. Alternatively, before this differentiation, a B cell can go to a GC, where it suffers somatic hypermutation to increase its affinity with the antigen. Plasma cells secrete antibodies, which bind to antigens. When an antigen is coated with antibody, it can be eliminated in several ways.

An explanation about cellular immune response follows. Some pathogens can escape antibody detection by infecting cells. Then, infected self-cells stimulate cytotoxic T cells, which activate a killing response. T cells must interact with helper T cells (using lymphokines) to regulate the destruction of infected cells. Overall processes of humoral and cell-mediated immune responses are illustrated in Figure 1.7.

When the immune system has been exposed to an antigen for a second time, it reacts quickly and rigorously (measured by the production of antibodies). This is called

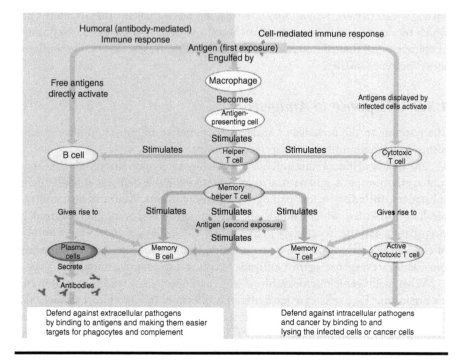

Figure 1.7 Illustration of humoral and cellular immunity.

secondary immune response, in contrast to the first encounter with the antigen, in which a slower response, called primary immune response, occurs (see Figure 1.8). This augmented antibody response is due to the existence of memory cells, which rapidly produce plasma cells on antigen stimulation. Thus, the immune system learns from encounters with antigens to improve its response in subsequent encounters, producing a so-called immunological memory.

1.3.3 T Cell Maturation

T cells are produced by the bone marrow and are initially inert, that is, they are not capable of performing their intended functions. To become immune-competent, they have to go through a maturation process. In B cells, the maturation process occurs in the bone marrow itself. T cells, instead, have to migrate to the thymus where they mature.

During maturation, T cells express a unique ABR on their surface called a T cell receptor (TCR). The generation of various TCRs is controlled by a random recombination of different gene segments. Unlike B cells, TCRs can only recognize antigenic peptides that are presented by cell-membrane (MHC) molecules (see Figure 1.9).

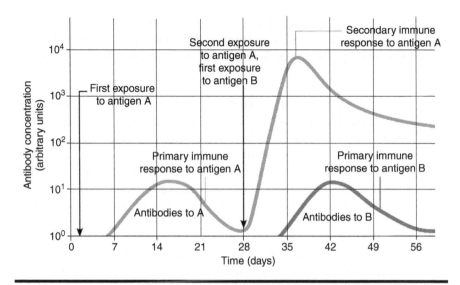

Figure 1.8   Immunological memory—primary and secondary immune responses.

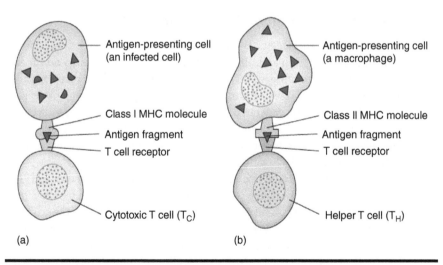

Figure 1.9   T cell recognition. T cells can only recognize antigen associated with MHC molecules on cell surfaces. (a) T cell recognition of class I MHC molecules, (b) T cell recognition of class II MHC molecules. (From Coutinho, A., *Ann. of Immunol. (Inst. Pasteur.)*, 131D, 1980; Paul, W. E., *Fundamental Immunology*. Raven Press, New York, 1993.)

During T cell maturation, they go through a process of selection, which ensures that they are able to recognize nonself peptides presented by MHC molecules. This process has two main phases: positive selection and negative selection (Coutinho, 1980; Paul, 1993).

- *Positive selection.* In positive selection, T cells are tested for recognition of MHC molecules expressed on the cortical epithelial cells. If a T cell fails to recognize any of the MHC molecules, it is discarded; otherwise, it is kept.
- *Negative selection.* The purpose of negative selection is to test for tolerant self-cells. T cells that recognize the combination of MHC and self-peptides fail this test. This process can be seen as a filtering of a diversity of T cells, in which only those that do not recognize self-peptides are kept (Kappler et al., 1987).

When a T cell encounters antigens associated with an MHC molecule on a cell, such a T cell will proliferate and differentiate into memory T cells and various effector T cells. Cellular immunity is accomplished by these generated effector T cells. There are different types of T cells that interact in a complex way to kill altered self-cells (for instance, virus-infected cells) or to activate phagocytic cells (Abbas and Lichtman, 2005; Moss et al., 1992).

1.3.4   B Cell Proliferation: Affinity Maturation

When receptors on the surface of a B cell bind to an antigen, this B cell gets stimulated to undergo proliferation and differentiation. Also, when receptors on the surface of a T cell bind to an antigen, such a T cell proliferates. This process is called clonal selection because antigen binding drives a particular cell for clonal expansion. Thereby, B cells that are generated become either memory cells or plasma cells. Memory cells ensure that subsequent infections by a pathogen receive a more rapid response, when plasma cells secrete large amounts of antigen-specific antibodies. Figure 1.10 illustrates B cell activation by specific antigens.

In early stages of the immune response, the affinity between antibodies and antigens may be low. But as B cells undergo clonal selection, they clone and mutate repetitively to improve the binding affinity between a particular antigen and a B cell type. This mutation process is called *somatic hypermutation*. Then, these activated B cells mature into plasma cells, which in turn produce antibodies with a high affinity of the Ag/Ab bonds. The entire process by which new B cells with high affinity to an antigen are created (clonal selection + somatic hypermutation) is called *affinity maturation*.

Ultimately, affinity maturation will lead to the production of a pool of antibody-secreting plasma cells and a pool of memory cells. Plasma cells are matured B cells that form a large endoplasmic reticulum for massively synthesizing and secreting specific antibodies. In contrast, memory cells are B lymphocytes with the same specificity receptors as those on the original activated B cell (Perelson and Oster, 1979).

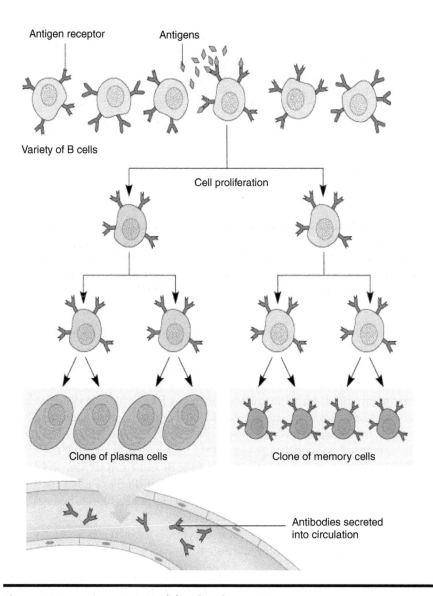

Antigen receptor

Antigens

Variety of B cells

Cell proliferation

Clone of plasma cells

Clone of memory cells

Antibodies secreted into circulation

Figure 1.10   Basic concepts of the clonal expansion.

1.3.4.1   Germinal Center

Lymph nodes are small nodular aggregates of lymphocyte-rich tissue situated along lymphatic channels throughout the body. A lymph node consists of an outer cortex and an inner medulla; and lymph nodes contain aggregates of cells called follicles, and evolve a specialized area called *germinal center* (GC).

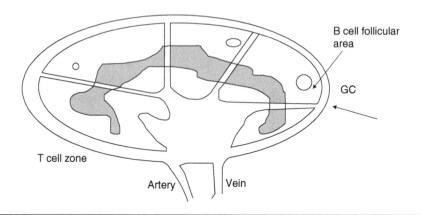

Figure 1.11 A diagram showing a section of a splenic lymph node.
(From Thorbecke, G. J. and V. K. Tsiagbe. *The Biology of Germinal Centers in Lymphoid Tissue*, Springer, Berlin, 1998; Liu, Y. J., G. Grouard, O. de Bouteiller, and J. Banchereau. *Int. Rev. Cytol.*, 166, 139–179, 1996.)

A GC is dynamically formed when antigen-activated B cells migrate into primary follicles of the peripheral lymphoid organs. However, the formation of GCs requires B cell activation and migration, T- and B cell interactions, and a network of follicular dendritic cells (FDC). A GC provides a specialized microenvironment to perform many critical immune functions such as a B cell's somatic hypermutation, clonal expansion, affinity maturation, and differentiation as memory or plasma cells. A GC is a biological immune system's functional module, which plays a major role in immune response (MacLennon, 1994; Todd and Spickett, 2005).

Figure 1.11 shows various zones where specific B- and T cell activities take place. The purpose of these activities is to generate a group of B cells that has the highest capability of recognizing a stimulating antigen (Liu et al., 1996; Thorbecke and Tsiagbe, 1998).

As shown in Figure 1.12, the number of centrocytes increases in the GC, two distinct regions begin to be distinguished:

1. The dark zone, where proliferation centroblasts are packed closely together and where there are few follicular dendritic cells, it is formed in a few days within a primary lymphoid follicle.
2. The light zone, in which centroblasts give rise to centrocytes that enter the follicular dendritic cells network, thereby, densely packed centrocytes make contact with the numerous cells of the follicular dendritic cell network.

Helper T cells that migrate to the primary follicle along with the activated B cells also undergo some clonal expansion and can be seen intermingled with centrocytes in the light zone. Centrocytes that fail to take up antigen from follicular dendtritic cells, die and are phagocytosed by local macrophages; also, cell death in the GC is

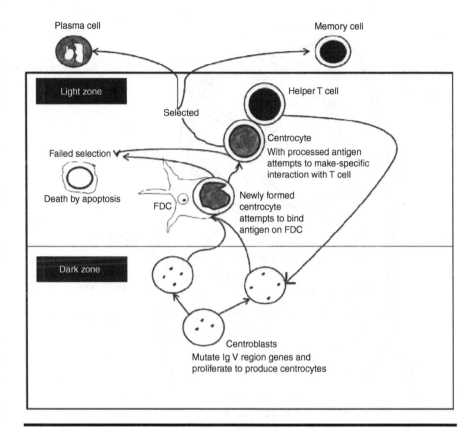

Figure 1.12   Different compartments of a GC such as light zone (selection chamber) and dark zone (for centroblasts).

seen to occur in the area of the light zone rich in follicular dendritic cells (Abbas and Lichtman, 2004; Hollowood and Goodlad, 1998). The exchange signals that induce further proliferation of the participating T and B cells, and differentiation of the latter results in producing either memory B cells or plasma cells (Hames and Glover, 1996; Harvard Medical School, 2005; Liu et al., 1996).

From an information-processing point of view, GCs can be viewed as production factories where highly specialized immune cells and molecules are evolved through an elegant recruitment process.

1.3.5   Apoptosis and Lysis

Apoptosis refers to programed cell death (cell suicide), the body's normal method of disposing damaged, unwanted, or unneeded cells. Lysis refers to the death of a cell by bursting, often by viral or osmotic mechanisms that compromise the integrity of the cellular membrane.

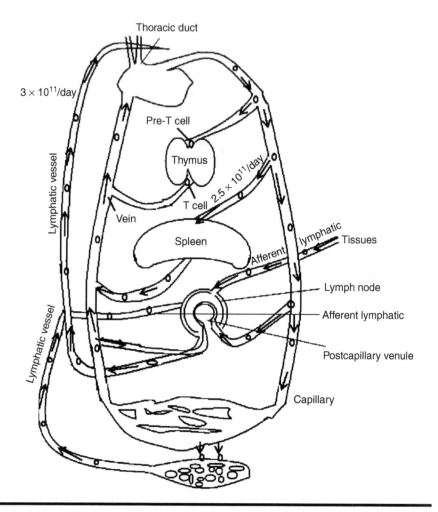

Figure 1.13   Circulation of immune cells through the blood (white) and lymph vasculature (black) to major organs of lymphatic systems. (From Kuby, J. et al., *Immunology*, 3rd edition. W. H. Freeman, New York, 2000.)

1.4   Circulatory Mechanism

As seen earlier, the immune system is an inherently distributed system that consists of a variety of specialized cells, enzymes, and other serum proteins, which are spread throughout the body. Immune cells, particularly, lymphocytes circulate constantly through the blood, lymph, lymphoid organs, and tissue spaces (Figure 1.13). They visit primary and secondary lymphoid organs to interact with foreign antigens. Studies show that lymphocytes circulate through blood for 2–12 hours before appearing in a particular lymphoid organ (Kuby et al., 2000).

This feature allows a maximum number of antigenically committed lymphocytes to encounter and interact with antigens within a relatively short period of time to generate a specific immune response. Different populations of lymphocytes circulate at primary and secondary lymphoid organs and are carefully controlled to ensure those appropriate B- and T-cell populations (naive, effector, and memory) are recruited into different locations. This differential migration of lymphocyte subpopulations at different locations of the body is called trafficking or homing.

As lymphocytes recirculate, they tend to home at various secondary lymphoid organs. Secondary lymphoid organs trap antigens and present them in pieces on the surface of APCs to be recognized by immune cells. These organs provide specialized environments to support clonal expansion and differentiation of antigen-activated lymphocytes into effector and memory cells. Interestingly, memory cells exhibit selective homing to the type of tissue in which they first encountered an antigen. Presumably, this ensures that a particular memory cell will return to the location where it is most likely to reencounter a subsequent antigenic challenge.

Experiments have shown that when a particular antigen is injected, antigen-specific T cells disappear from circulation within 48 hours. This suggests that specific T cells encounter an antigen in peripheral lymph organs and cease recirculating within such a time period. This process is closely regulated to guarantee steady-state levels of each blood cell type. Cell division and differentiation of each lineage is balanced by programed cell death known as apoptosis. Such a behavior provides increased sophistication without centralized control.

1.5 Regulatory Mechanisms

Immune response mechanisms are self-regulatory by nature. There is no central organ that controls immune functions. The regulation of immune responses can be broadly divided into two branches: humoral immunity, mediated by B cells and their products and cellular immunity, mediated by T cells. Both branches follow a similar sequence of defense steps: proliferation, activation, induction, differentiation and secretion, attack, suppression, and memory; however, they do it in different ways.

When an antigen enters the body, self-regulatory mechanisms determine (influenced by prior exposure to antigen) the branch of the immune system to be activated, the intensity of the response, and its duration. Specifically, regulation of both humoral and cellular immunity is conducted by a population of T cells referred to as either helper or suppressor cells, which either augment or suppress immune responses. T cells regulate immune responses by releasing soluble molecules, collectively referred to as cytokines to activate B cells. Subsequently, B cells follow one of two pathways: they either differentiate into plasma cells, which are basically antibody-secreting factories, or they give rise to GCs, specialized structures within lymphoid organs, where they undergo somatic mutation (a process called affinity maturation).

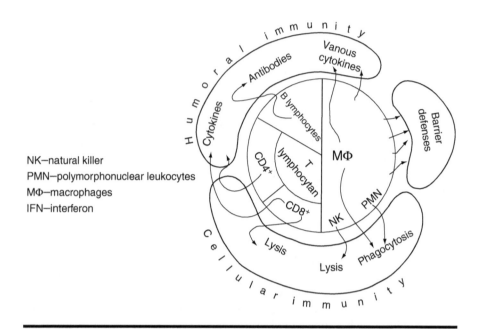

NK—natural killer
PMN—polymorphonuclear leukocytes
MΦ—macrophages
IFN—interferon

Figure 1.14 Overall immunity (coverage) with different defense mechanisms. (From Whitton, J. L., *Curr. Top. Microbiol. Immunol.*, 232, 1–14, 1998.)

The importance of self-regulatory mechanisms is evident in clonal expansion (in humoral immunity) due to the presence of an antigen; this is also observed when specific immune cells are reduced after the clearance of antigens. Such a control of antibody production is thought of as an idiotypic regulatory network (Jerne, 1974).

Moreover, B cell clonal expansion and proliferation are closely regulated to prevent uncontrolled immune response. This second signal helps to ensure tolerance and discrimination between dangerous and harmless invaders. So the purpose of this accompanying signal in identifying nonself is to minimize false alarms and to generate a decisive response in case of a real danger. Figure 1.14 depicts defense strategies (outer layers), involvement of immune cells and molecules (inner circle), and their interconnecting mechanisms in the biological defense system (Whitton, 1998).

1.6 Signaling and Message-Passing Mechanism

In the immune system, signal diffusion and dialogue are noticeable as two kinds of communication schemes. They play a major role in sharing and passing information during the immune response. In *immune diffusion*, the message is passed from one immunocomponent to others without any feedback. Another scheme is called *immune dialogue*, where the immune system components continuously exchange molecular signals with their counterparts. Immune sensitivity is determined by context, where an immune cell and a pathogen play on one another. The body is under constant

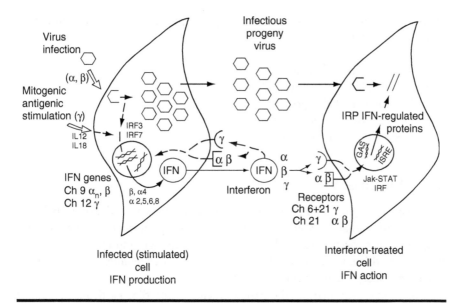

Figure 1.15 Illustration of IFN signaling mechanism.

challenge from a continuum of hostile behaviors and needs to respond judiciously through a coordinated decision process (Tew et al., 1997).

The protein interferon (IFN) is produced by cells when they are invaded by viruses; it is released into the bloodstream or intercellular fluid to induce healthy cells to manufacture an enzyme that counters the infection. Figure 1.15 illustrates the release of IFN by infected cells during virus infection. Since the receptor for IFN has a common structure among different host cells, the IFN that is produced by a virus-infected cell can bind to a receptor of the neighboring cell and enters it. IFN induces the production of IFN-regulated proteins within the neighboring cell; such proteins inhibit the virus replication inside the cell.

The following is a list of reasons why signaling is important in biological defense:

■ It allows a cell to move a signal from outside to inside.
■ Signaling results in changes to the cell, allowing it to appropriately respond to a stimulus.
■ It allows signaling and message passing among various functional components.
■ It allows response to external stimuli such as cytokines, growth factors, hormones, tissue repair or remodeling, and stress.
■ Tissue-specific regulation is the hallmark of virus-driven cytotoxic T cells expansion in immune response.
■ Signaling regulates differentiation and development, and immune response.

Additionally, cells communicate with surroundings through their surface receptors that recognize extracellular signal and convert into intracellular signal, which then transmit toward the nucleus to develop proper response. This cellular communication process results in

- Surface marker changes
- Changes in cellular distribution
- Environmental changes
- Destruction of foreign invaders
- Destruction of anomalous cells

1.7 Summary

This chapter is intended for the readers having no or little knowledge of the BIS, and discusses topics that are available in basic immunology textbooks. However, this chapter neither covers latest findings nor provides enough information for immunologists as this book is primarily intended for researchers interested in abstract biological concepts and developing artifacts.

This chapter summarizes the basic elements of the BIS. Particularly, roles of various organs, immune cells (such as B cells, T cells, and other lymphocytes), and the overall process by which an antigen recognition occurs are described. Thus, the purpose of this chapter is to provide an abstract view of the biological immune system and its important mechanism that inspired to develop computational models.

1.8 Review Questions

1. Write short notes on antigen, antibody, innate immunity, and adaptive immunity.
2. What is the role of B cells in immune response?
3. What is the role of T cells in immune response?
4. What is the role of helper T cells?
5. What are the main differences between humoral immunity and cell-mediated immunity?
6. What is the main distinction between self/nonself recognition?
7. What is the role of APCs in immune response?
8. Explain the statement "antigen recognition is approximate."
9. How is the circulatory system involved in immune response?
10. What is the purpose of immune memory?
11. What is the relationship between antibodies and B cells?
12. Explain the terms clonal selection, somatic hypermutation, and affinity maturation.
13. Describe with a diagram the clonal expansion process.
14. Which are the immune system's features that make it inherently distributed and fault-tolerant?

15. Compare the roles of B- and T cells in immune response.
16. What are the two selection processes that are used in T cell maturation?
17. What is trafficking (or homing)?
18. What is a GC? What is the role of a GC in the immune system?
19. Under what conditions does a GC start forming in a lymph node?
20. What happens when the number of centrocycles increases in the GC? Define.
21. Which are the two stages involved in the centrocycle? Explain briefly.
22. What happens when a high rate of mutation couples with clonal competition? Discuss its effects.
23. Match immune concepts in column A with their corresponding concepts in column B.

Immune System Structure	
Column A	Column B
Bone marrow	Direct the cellular-mediated immune response and phagocytosis
Hematopoiesis	Its function is to produce mature T cells
Thymus	Formation of blood cells process
Spleen	Specialized cells, which engulf large particles by a process called phagocytosis
Lymph nodes	Organs where all the immune system's cells are initially derived from
Cytokines	It consists of an outer cortex and an inner medulla
Helper T cells	Antigen-binding proteins present on a B cell membrane
Antibodies	Aid B cells to perform antigen recognition by releasing cytokines
MHC proteins	It is an immunologic filter of the blood
Macrophages	They act as "sign posts" that display peptides on the surface of a host cell
Complement system	System of molecules that leads a sequence of events on the surface of a pathogen that helps destroy it

24. Complete the missing words (fill in the blanks) in the following diagram.

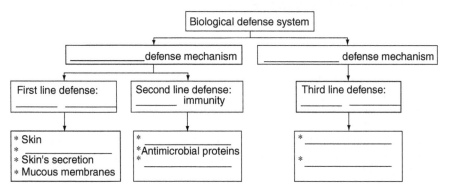

25. What are the roles of MHC molecules in recognizing antigenic peptides?
26. Which are the mechanisms that define the immune system's dynamical behavior?
27. Why does the immune system react in a stronger and faster fashion when it is exposed to an antigen for a second time?
28. Explain the functional differences between ABRs in T and B cells during maturation.
29. Explain positive and negative selection during T cell maturation.
30. Order the following list of processes with respect to their temporal occurrence in B cell activation. Explain each process.

    • An antibody on the surface of a B cell binds to an antigen.

    _____

    _____

    • Clonal selection

    • Antibodies become either memory cells or plasma cells

    _____

    _____

    • Somatic hypermutation

    _____

    • Affinity maturation

    _____

    • Production of a pool of antibody-secreting plasma cells and memory cells

    _____

    _____

31. How is the circulatory system involved in immune response?
32. Do humoral and cellular immunity follow a similar sequence of defense steps?
33. Explain the difference between immune diffusion and immune dialogue.
34. Why is signaling important in biological defense?
35. Draw a diagram showing different strategies and mechanisms providing overall immune defense (coverage).

References

Abbas, A. K. and A. H. Lichtman. *Basic Immunology: Functions and Disorders of the Immune System*, Saunders, 2004.

Abbas, A. K. and A. H. Lichtman. *Cellular and Molecular Immunology*, 5th Edition, Saunders, 2005.

Baumann, H. and J. Gauldie. The acute response. *Immunol Today*, 15(74), 74–80, 1994.

Clancy, J. *Basic Concepts in Immunology: A Student's Survival Guide*, McGraw-Hill, New York, 1998.

Coutinho, A. The self non-self discrimination and the nature and acquisition of the antibody repertoire. *Ann. Immunol. (Inst. Pasteur.)*, 131D, 235–253, 1980.

Hames, B. D. and D. M. Glover. *Molecular Immunology*, Oxford University Press, Oxford, 1996.

Harvard Medical School. *The Truth About Your Immune System: What You Need to Know*, Harvard Health Publications, 2005.

Hollowood, K. and J. R. Goodlad. Germinal centre cell kinetics. *J. Pathol.*, 185(3), 229–233, 1998.

Jerne, N. K. Towards a network theory of the immune system. *Ann. Immunol. (Inst. Pasteur.)*, 125C, 373, 1974.

Kappler, J., N. Roehm and P. Marrack. T cell tolerance by clonal elimination in the thymus. *Cell*, 49(2), 273–280, 1987.

Kuby, J. (Ed.) *Immunology*, 3rd Edition, W. H. Freeman, New York, 2000.

Liu, Y. J., G. Grouard, O. de Bouteiller and J. Banchereau. Follicular dendritic cells and germinal center. *Int. Rev. Cytol.*, 166, 139–179, 1996.

MacLennon, I. C. M. Germinal center. *Annu. Rev. Immunol.*, 12, 117, 1994.

Moss, P. A. H., W. M. C. Rosenberg and J. I. Bell. The human T cell receptor in health and disease. *Annu. Rev. Immunol.*, 10, 71–96, 1992.

Pathak, S. and U. Palan. *Immunology: Essential and Fundamental*, 2nd Edition, Science Publishers, Enfield, NH, 2005.

Paul, W. E. *Fundamental Immunology*, Raven Press, New York, 1993.

Perelson, A. S. and G. F. Oster. Theoretical studies of clonal selection: Minimal antibody repertoire size and reliability of self- non-self discrimination. *J. Theor. Biol.*, 81(4), 645–670, 1979.

Sompayrac, L. M. *How the Immune System Works*, 2nd Edition, Blackwell Publishing, Oxford, 2003.

Stanley, J. *Essentials of Immunology and Serology*, Thomson Delmar Learning, Australia, 2002.

Tew, J. G., J. Wu, D. Qin, S. Helm, G. F. Burton and A. K. Szakal. Follicular dendritic cells and presentation of antigen and costimulatory signals to B cells. *Immunol. Rev.*, 156, 39–52, 1997.

Thorbecke, G. J. and V. K. Tsiagbe. *The Biology of Germinal Centers in Lymphoid Tissue*, Springer, Berlin, 1998.

Todd, I. and G. Spickett. *Lecture Notes: Immunology*, Blackwell Publishing, Oxford, 2005.

Whitton, J. L. An overview of antigen presentation and its central role in the immune response. *Curr. Top. Microbiol. Immunol.*, 232, 1–14, 1998.

Woods, S. L. *Understanding the Immune System. U.S.: Department of Health and Human Services*, Public Health Service, National Institutes of Health, 1991.

Chapter 2

Theoretical Models of Immune Processes

A feature of the adaptive immune response is its ability to "remember" its encounters with antigens to achieve a faster response when the same antigen is confronted at a later time. However, the mechanisms involved in this behavior are not fully understood. There are several theories in the literature about how this behavior is achieved (Burnet, 1959; Celada and Seiden, 1996; Matzinger, 2002; Oprea and Perelson, 1997; Varela and Stewart, 1990).

One hypothesis states that B lymphocytes that have reacted to an antigen simply remain in a dormant state for years, waiting for a recurrence of the same antigen. It is known that B lymphocytes can remain in a memory state for periods of weeks or possibly months; however, it is not known if they can survive for years without being stimulated. Some immunologists think that a kind of internal restimulation keep this immune memory preserved for a long time. Another hypothesis suggests that the antigen itself (or some partially degraded form) is sequestered in lymph nodes and other organs and stimulate the immune system through periodical exposure to the antigen, thereby reinforcing the memory.

In the following sections, some works intended to model various immunological principles and mechanisms are described to better understand the biological processes and simulate its dynamic behavior during the immune response.

2.1 Clonal Selection Theory

Burnet (1959) proposed "clonal selection theory" to explain the proliferation of immune cells in the presence of an antigen. This theory states that an antigen

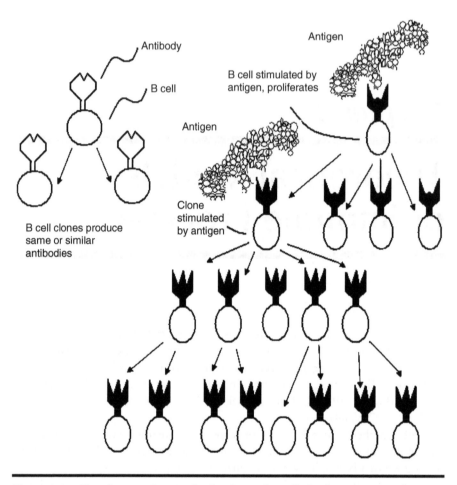

Figure 2.1 Clonal expansion (and selection) of B cells in the presence of an antigen.

selects (or induces) a particular lymphocyte (to produce clones) from a large number of lymphocytes. Given the large variety of lymphocyte receptors in the expressed repertoire, whether the receptors on any particular lymphocyte detects the antigen can be viewed as a random event. On activation, however, this lymphocyte proliferates through the process called cloning. Figure 2.1 illustrates clonal expansion and selection of B lymphocytes in the presence of an antigen (Hightower, 1996).

The main properties of clonal selection are (Burnet, 1976; Stewart et al., 1989)

■ Elimination of self-reacting clones
■ Proliferation and differentiation of mature lymphocytes through antigenic simulation

- Restriction of one pattern to one differentiated cell and retention of this pattern by clonal descendants
- Generation of new random genetic changes subsequently expressed as diverse antibody patterns by a form of accelerated somatic mutation

Other researchers also investigated the clonal selection to understand how certain types of B and T lymphocytes are selected for destruction of specific antigens invading our body.

2.2 Immune Network Theory

The "immune network (IN) theory" was developed by Niels K. Jerne (Jerne, 1974) in an attempt to explain how the immune memory gets formed. He hypothesized that the immune system acts as a regulated network of antibodies and anti-antibodies, called an "idiotypic network," which recognizes one another (even in the absence of antigens) rather than being a set of isolated clones (antibodies of the same specificity) that respond only when stimulated by antigens. According to Jerne, B and T cells form a complex circuitry of interacting cells that functions either to stimulate or to suppress the immune activation. Although there is evidence of the existence of the idiotypic network, its physiological relevance has been much debated. Moreover, the complexity of idiotypic networks has made it difficult to predict whether administration of anti-idiotype antibodies or T cells bearing anti-idiotype receptors up- or down-regulate immune responsiveness (Kuby et al., 2000).

Mature B lymphocytes carry highly specific receptors (antibodies) on their surface. These receptors are stimulated by complementary structures and such stimulation causes proliferation of particular antibodies (Figure 2.2). The portion on an antigen's surface that an antibody is able to recognize is called an epitope, and the corresponding part of an antibody used to recognize antigens is called a paratope; an antibody's epitope is called an idiotype (see Figure 2.3).

According to Jerne's theory, a sequence of events forms an idiotypic network as follows. First, an antigen is recognized by B cells, which secrete antibodies Ab1. Ab1 antibodies themselves are also recognized by "anti-idiotypic" B cells that secrete antibodies Ab2. Thus, further interactions can lead to Ab3 antibodies that recognize Ab2 and so on. In an idiotypic network, there is no intrinsic difference between an antigen and antibody; and any node of the network can bind to and be bound to by any other.

Several IN models have been proposed, which can be classified into three generations, where each version of IN incorporates additional features of immune processes.

2.2.1 First-Generation Immune Networks

First generation IN (FGIN) model tries to predict the number of different types of antibodies (clone) present in the blood. The increase or decrease of the number of

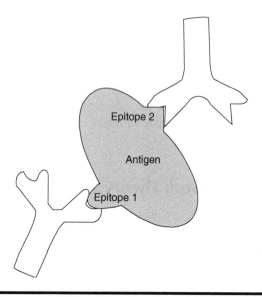

Figure 2.2 **The portion of an antigen that is recognized by an antibody is called an epitope, and an antigen may have multiple epitopes.**

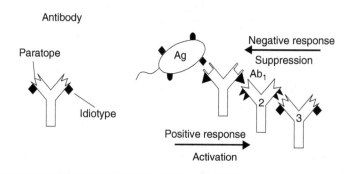

Figure 2.3 **Antibodies are recognized by anti-antibodies by matching their idiotopes to another antibody's paratopes.**

clones depends on its interactions with idiotypically related clones and the antigen; thereby, the increased number of each clone determines the strength of immune response. In this aspect, FGIN theory competes with clonal selection theory.

FGIN models concentrate on the interactions among free antibody molecules, giving little or no attention to the biology of the cells that produced such antibodies; thus, the internal state of B cells and its interaction with other B and T cells is not considered (Coutinho, 1989).

2.2.2 Second-Generation Immune Networks

The second-generation IN (SGIN) refers to a specific class of networks that tries to model natural autonomous behavior of the immune system (Varela and Coutinho, 1991). Experimental and theoretical studies on naturally activated T and B cells (cells activated without the presence of antigens) shows that natural serum antibodies and activated B cells influence other (activated B and T) cells, whereas resting B and T cells do not have such effects. The SGIN model postulates that interactions of B cells in the network are determined by their affinity to preexisting soluble antibodies and self-antigens and have distinct functional consequences, according to a bell-shaped dose response curve, which empirically determined it (Varela and Coutinho, 1991). Thus, both very low and very high levels of interactions lead to cell death, whereas intermediate levels of interaction (for increased binding strength) result in

1. Cell survival in the resting state
2. Cell minimal proliferation
3. Cell proliferation with little or no antibody secretion

Therefore, SGINs try to model the immune system's autonomous behavior, that is, its behavior in the absence of antigens. Also, these models involve the notion of network "dynamics" and metadynamics, which include the rate of production of new B cells, turnover rates in the resting lymphocyte compartment, and the rates of their activation and differentiation to antibody-secreting plasma cells.

The main contribution of SGIN model is that it brought two views of the immune system into compatibility, clonal selection and network theories (Coutinho, 1989). Although the clonal selection theory tries to explain the specificity and amplification of immune responses to external antigens, this IN theory looks for explanations to the way "preimmune" repertoires are selected, how natural lymphocytes are activated, self-tolerance, and the biology of autoreactive cells.

To summarize, the aspects of SGIN models are as follows:

- INs are made up of B lymphocyte clones, which are connected through idiotypic interactions. Interactions of these INs with T lymphocytes are neglected or not considered.
- The strength of activation and population dynamics of each lymphocyte clone is controlled by the strength of receptor ligation in soluble Ig molecules.
- Soluble Ig molecules are the main mediators of idiotypic interactions because they can rapidly diffuse through the body fluids, in a much higher proportion than those present as membrane Ig receptors in the lymphocytes.

Other versions of SGIN models (De Boer, 1989; Neumann, 1992; Wiesbuch et al., 1990) do not distinguish between free and cell membrane–bound immunoglobulins; however, it has shown that such assumption does not alter the main conclusions of SGIN models.

2.2.3 Third-Generation Immune Networks

As a result of criticizing SGIN models and questioning the assumption that T cell help is never a limiting factor for B-lymphocyte proliferation or antibody production, Stewart and Carneiro (1999) proposed an extended version of the model (proposed by Varela et al., 1988), which is known as third generation INs (TGIN).

TGINs introduced the concepts of the central immune system (CIS) and peripheral immune system (PIS). The CIS represents a group of activated, autoreactive, and interconnected lymphocytes, which represents 10–15 percent of the total number of lymphocytes. The PIS contains the remaining 85–90 percent of all type of lymphocytes that are encountered in lymphoid organs (Stewart and Carnerio, 1999).

The CIS is composed of a network of clones, which exhibits autonomous activity and integrates antigens into its ongoing regulatory dynamics. However, the PIS is composed of lymphocyte clones, which remain in a resting state unless they are specifically activated by an antigen resulting in immune response. Thus, the PIS only represents potential network targets for eventual recruitment through activation, in case this network space evolves to include them. Therefore, PIS takes care of reactions with the immunizing antigens. In contrast, CIS deals with body antigens. It is also assumed that resting lymphocytes are disconnected from regulatory influence from the network, thus providing them ideal conditions to respond to external antigens. Therefore, conventional antigens provide specific stimulation according to Burnet's clonal selection theory (Burnet, 1959).

TGIN models incorporate B and T cell cooperation to accommodate both structural and functional properties of CIS and PIS in a coherent way, and also explain how the CIS and PIS distinction can emerge from the self-organizing properties of the network. To define the frontier between CIS and PIS, the TGIN model (Stewart and Carneiro, 1999) assumed that any given lymphocyte clone belongs either to the CIS or to the PIS at any given time; they showed distinction between the structure and function and exhibit cooperation between B and T lymphocytes. New antibodies produced through antigenic experiences enter the network and alter its organization, thus allowing the formation of a "systemic memory." However, expanded clones find their internal legends in the network structure for their long-term preservation, even in the absence of external antigens.

In TGINs, the B cell activation is considered to occur in two steps:

- *Induction.* The activation of lymphocytes based on the degree of stimulation on a receptor by cross-linking agents such as anti-idiotypic circulating antibodies or self-antigens. In this model, the induction is described by a characteristic lognormal function, as in SGINs.
- *Cooperation between an induced B cell and activated T cell.* This cooperation leads to the "full" activation of the B cell. Thus, a function that describes the competition of B cells to get help from T cells is defined as a function of

the B or T cell affinities over the total set of B or T cells. An induced B cell cooperates with an activated T cell if it engages its T cell receptor by either acting as an antigen-presenting cell (APC) or anti-antibody (anti-idiotypic interaction).

Dynamic behavior of T cell clones is similar to that of B cell clones; however, T cell clones are only driven by antigenic peptides on APCs. Specific peptides from antibodies expressed and produced by B lymphocytes are not considered due to their low individual concentration and frequency. Thereby, a bounded dynamics of T cell clones is attained if and only if their receptors are considered a part of the idiotypic network.

2.3   Multiepitope Immune Network

This approach tried to map the IN theory into a parallel distributed processing (PDP) (Vertosick and Kelly, 1989). They argued that B lymphocytes (or lymphocyte clones) can act as the units that compose a PDP network, that is,

- Receive inputs (from APCs, antigens, and cytokines)
- Generate output (antibody)
- Remember antigenic specificity
- Convert inputs (antigenic stimulation) into output (antibody secretion) in a quantitative fashion

The PDP IN architecture can be designed to be multilayered, where lymphocytes, plasma cells, and the lymphocytes that produce anti-idiotypic antibodies are considered as input, output, and hidden units, respectively (Figure 2.4).

The connection weights between two lymphocytes can be defined in terms of its affinities toward one another. The learning behavior of the immune system uses an unsupervised, local (Hebbian) learning rule. This model also includes cytokines, which are responsible for the clonal expansion of the population and subsequent alteration of the connection strengths of a PDP composed of clonal units. The simulated annealing technique is used to find the lowest energy configuration of the PDP network, by altering the shape of the activation functions of the units. Using this model, complex antigen patterns (consisting of multiple epitopes) can be learned and stored within the network.

2.4   Modeling the Germinal Center

Germinal centers (GCs) are the sites (in the follicles of the secondary lymphoid nodes) where antigen-stimulated B cells complete their affinity maturation process (Berek et al., 1991; Casamayor-Palleja et al., 1997; MacLennan). Particularly, GC

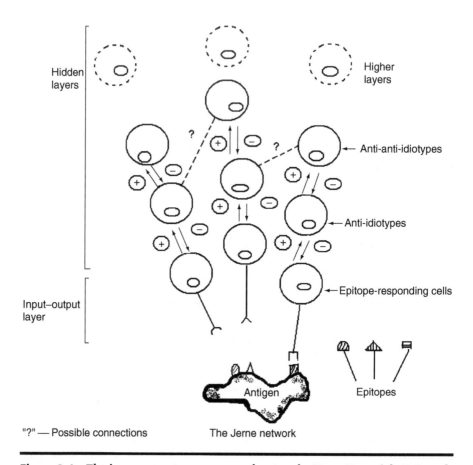

Figure 2.4 **The immune system as a neural network. (From Vertosick, F. T. and R. H. Kelly, *J. Theoretical Biol.,* 150, 225–237, 1991; Vertosick, F. T., and R. H. Kelly, *Immunology,* 66, 1–7, 1989)**

performs many critical immunological functions by providing a specialized micro-environment for proliferation of B cells through clonal expansion and somatic hypermutations (Celada and Seiden, 1996; Jacob et al., 1991; Kepler and Perelson, 1993; Leanderson et al., 1986; Liu et al., 1989), giving rise to plasma and memory B cells. Details of GC formation are covered in Chapter 1.

The GC initially contains only dividing centroblasts (Camacho et al., 1998), but shortly evolves into dark and light zones (Liu et al., 1989). Figure 2.5 shows the GC mechanisms, where the dark and light zones are involved in clonal expansion and clonal selection, respectively.

There are several mathematical models (differential equations) developed to simulate the GC dynamics (Kesmir and Boer, 1999). For example, the OP model

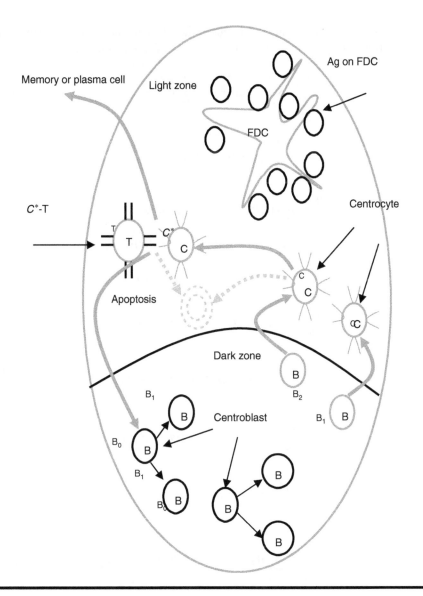

Figure 2.5   Illustration of different processes in a GC.

(Oprea and Perelson, 1997) assumed that the dark zone of GC is mainly for proliferation of centroblasts and no selection of centrocytes occurs at this stage. When the number of centroblasts reaches a certain number, proliferation stops and selection of centrocytes begins. The selected centrocytes are used for antigen (Ag) binding with the follicular dendritic cell (FDC) network and thus form

Ag complexes. The involvement of T cells in the GC dynamics is not included in this model. The set of ordinary differential equations used in this model are as follows:

$$S(t) = S_0 e^{-dot} - \sum_{i=-2}^{3} [C_i - S] \tag{2.1}$$

$$\frac{dB}{dt} = p_B B \left(1 - \frac{B}{M}\right) - k_d \phi(t) B \tag{2.2}$$

$$\frac{dB_i}{dt} = k_d \phi(t) B \delta_{i,0} + 2 p_{cb} [\tau(i, i) B_i \theta_i$$
$$+ \tau(i-1, i) B_{i-1} \theta_{i-1} + \tau(i+1, i) B_{i+1} \theta_{i+1}]$$
$$+ p_r m_R R_i - f_m B_i - d_B B_i - p_{cb} B_i \tag{2.3}$$

$$\frac{dC_i}{dt} = f_m B_i - k_i^f C_i S + (1 - \rho) k_r [C_i - S] \tag{2.4}$$

$$\frac{d[C_i - S]}{dt} = k_i^f C_i S - k_r [C_i - S] \tag{2.5}$$

$$\frac{dR_i}{dt} = \rho k_r [C_i - S] - (m_R + d_R) R_i \tag{2.6}$$

$$\frac{dM_i}{dt} = (1 - p_r) \eta m_R R_i - d_M M_i \tag{2.7}$$

$$\frac{dL_b}{dt} = 2 p_{cb} \left[\tau_L \sum_{j=-2}^{3} \theta_j B_j + \tau(-2, -3) \theta_{-2} B_{-2} \right] - f_m L_B - \alpha L_B \tag{2.8}$$

$$\frac{dL_c}{dt} = f_M L_B = \alpha L_C \tag{2.9}$$

where S is the number of free FDC sites, S_0 the total initial number of FDC sites, $\phi(t) = t^n/k^n + t^n$, $k = 6$, $n = 40$, i the affinity class, B_i the centroblasts, C_i the centrocytes, $[C_i - S]$ the complex of centrocytes with FDC, R_i the rescued centrocytes, M_i the memory, m_R the migration rate of the respective cells, p_{cb} the constant centroblast proliferation rate, d_B the death rate, L_B the centroblasts that are not in the affinity class, αL_B the rate at which L_B die, k_r the reverse rate constant, R_i the rescued centrocytes that migrate out of GC with rate $m_R R_i$, and η a fraction that enters the memory compartment.

Although the Kesmir–Boer model (Kesmir and Boer, 1999) includes the T cell dynamics, it contradicts with the OP model because it allows proliferation and selection almost simultaneously and allows the proliferation of centrocyte complex (C^*) and T cells after centrocyte selection. The set of differential equations used in Kesmir–Boer model is as follows:

$$\frac{dB_0}{dt} = P_r \mu C_T^* - \rho B_0 + \delta_B B_0 \tag{2.10}$$

$$\frac{dB_i}{dt} = \rho(1 + \alpha_B) B_{i-1} - \rho B_i - \delta_B B_i \tag{2.11}$$

$$\frac{dC}{dt} = d \rho B_n - \mu C \tag{2.12}$$

$$\frac{dC^*}{dt} = \mu C_A - \mu C^* \tag{2.13}$$

$$\frac{dM}{dT} = (1 - p_r) \mu C_T^* \tag{2.14}$$

$$\frac{dA}{dT} = -zA - u C_A \tag{2.15}$$

$$\frac{dT}{dT} = \sigma + p\alpha_T C_T^* - d_T T \tag{2.16}$$

where $C_A = (C.A)/(S + A)$, S is a saturation constant, P the C^* recycle probability, μ the C disappear rate, ρ the B cell division rate, δ_B the B cell death rate, δ_T the T cell death rate, d the B to C phenotype conversion rate, z the A decay rate, u the C to C^* uptake rate, σ the initial T influx, and p the C^* proliferation rate.

Another recent model called the Dasgupta, Kozma, and Pramanik (DKP) model modified the Kesmir–Boer model and simulated the GC dynamic with a cascade of three Hopfield neural networks. It assumed that the migrated centroblasts are selected by Ag and T cells for forming a complex within the FDC network. This complex may dissociate into memory or plasma cells and a centrocyte complex (C^*) that feedback to the proliferation chamber for further production of high-affinity centroblasts.

In DKP GC simulation model, there are three basic chambers (proliferation, selection, and memory), each providing the excitatory and inhibitory (optional) layers to allow nontrivial oscillatory dynamics in each GC subunit (Pramanik et al., 2002).

2.5 Danger Theory

There has been a long-standing debate among immunologists on the validity of the classical self/nonself (SNS) discrimination theory (Bretscher and Cohn, 1970; Hoffmann, 1975) and its importance in the detection and recognition processes. Some alternative views have been proposed, such as the danger theory (DT) (Matzinger, 1994), integrity model (Dembic, 2000), and the self-assertion model (Varela and Coutinho, 1991).

The self-assertion model considers the danger as a result of the interaction between external stimulus and current state of the immune system; also, the same stimulus can produce different responses at different times (Bersini, 2002). According to this approach, the reaction to an antigen will depend on the system's evolution, which also includes the history of all previous external stimuli; thereby, it is assumed that an organism is in a sate of homeostasis—some metabolic equilibrium actively maintained by complex biological mechanisms that operate through autonomous behavior to counterbalance disrupting changes. This approach suggests that more attention should be devoted to self-regulation mechanisms, which allow the immune system to maintain a viable organization, despite the presence of numerous different dynamic elements with complex behaviors (Varela and Coutinho, 1991).

The DT (Matzinger, 1994), which is claimed to be a more realistic approach compared to the classical SNS discrimination model (Von Boehmer and Kisielow, 1990; Bretscher and Cohn, 1970; Hoffmann, 1975), is discussed next. In an SNS discrimination model, the word "foreign" is used to mean "that to which the immune system should respond," whereas the term "self" means "that to which the immune system should be tolerant." In DT model, the self is considered as harmless elements to which the body develops tolerance, which may also include antigens such as commensal organisms and parasites that do not cause any cellular damage. Accordingly, the immune system becomes tolerant to antigens that are not necessarily the same as self, which do not pose any danger (Figure 2.6).

In DT, the immune response is determined by the presence or absence of alarm signals; some "danger signals" such as tissue damage triggers a myriad of immune reactions and responses and APCs are activated by endogenous cellular alarm signals from distressed or injured cells.

According to Matzinger (1994), our bodies are never completely tolerant. As long as the thymus and bone marrow are producing new B and T cells, there will be a few new circulating self-reactive lymphocytes. Thus, a question that arises is "what stimulates these self-reactive cells and what maintains the response."

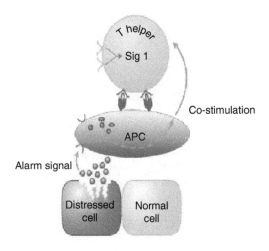

Figure 2.6 Illustration of the DT model, which includes alarm signals as a part of immune response.

She considers that there are, at least, four different categories of stimulus, which cause different types of autoimmune diseases.

1. *An unrecognized infection in the target tissue.* This is not actually an autoimmune disease, because self-antigens are not the primary target. In this case, although the immune system attempts to eliminate a pathogen, it ends up damaging the target tissue.

2. *Molecular mimicry by a pathogen that has some similarity to a self-tissue.* Some pathogen-specific T cells also detect self-antigens and then, they respond against both the pathogen and the self-tissues.

3. *Bad death.* This category is only considered in the DT model, as there are no infectious agents and foreign components. Cells die permanently in our bodies, and such deaths are controlled by some genes, which are subject to mutations as all genes do. However, such mutations could induce alarm signals to initiate immune responses. Also, environmental toxins that cause cell damage could lead to the release of such alarm signals.

4. *The wrong class.* As antibodies and cytokines released during an immune response are potent molecules that are intended to eliminate pathogens, they can damage some tissues, which are more sensitive to certain effector molecules than others. In this case, the immune system can kill a tissue with the wrong class of immunity, although the response is not intended to attack that particular tissue.

Figure 2.7 shows an abstract view of the antigen universe and illustrates the partitions considered in classical SNS discrimination model and DT. According to the

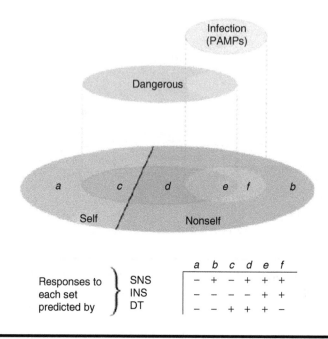

Figure 2.7   An abstract view of the antigen universe based on SNS, INS, and DT models.

SNS model, the self-set consists of subsets *a* and *c*; thus, self-elements in *a* and *c* are labeled as "−", whereas nonself subsets *b*, *d*, *e*, and *f* are labeled as "+". However, according to the infectious nonself model (INS), only part of nonself is considered infectious; thereby, nonself subsets *e* and *f* are labeled as "+". However, in DT (Matzinger, 2002), a subset of self is also considered to trigger alarm signals. Therefore, self-subset *c* and nonself subsets *d* and *e* are labeled as "+".

2.6   Computational Aspects of the Immune System

From the point of view of information processing, the biological immune system exhibits many interesting characteristics; some of which are (Dasgupta, 1999)

- *Pattern matching.* The immune system is able to recognize specific antigens and generate appropriate responses. This is accomplished by a recognition mechanism based on chemical binding of receptors and antigens. This binding depends on their molecular shapes and electrostatic charges.
- *Feature extraction.* Generally, immune receptors do not bind to a complete antigen, but rather to portions of it (peptides). Accordingly, the immune system can recognize an antigen just by matching segments of it. Peptides are

presented to lymphocyte receptors by APCs. Therefore, such APCs act as filters that can extract the important information and remove the molecular noise.

- *Learning and memory.* A major feature of the adaptive immune system is that it is able to learn through its interaction with the environment. The first time an antigen is detected, a primary response is induced, which includes proliferation and subsequent reduction of lymphocytes. Some of these lymphocytes are kept as memory cells. The next time the same antigen is detected, memory cells generate a faster and more intense response (secondary response). Accordingly, memory cells work as an associative (highly) distributed memory.

- *Diversity.* Clonal selection and hypermutation mechanisms are constantly testing different detector configuration for known and unknown antigens. This is a highly combinatorial process that explores the space of possible configurations looking for close-to-optimum receptors that can cope with all types of antigens. Exploration is balanced with exploitation by favoring the reproduction of promising individuals.

- *Distributed processing.* Unlike the nervous system, the immune system is not centrally controlled. Detection and response can be executed locally and immediately without communicating with any central organ. This distributed behavior is accomplished by billions of immune molecules and cells that circulate around the blood and lymph systems and are capable of making decisions in a local collaborative environment.

- *Self-regulation.* Depending on the severity of the attack, response of the immune system can range from very light and almost imperceptible to very strong. A stronger response uses a lot of resources to help ward off the attacker. Once the invader is eliminated, the immune system regulates itself to stop the delivery of new resources and release the used ones.

- *Self-protection.* By protecting the body as a whole, the immune system is also protecting itself. It means that there is no other additional system to protect the immune system; hence, it can be said that the immune system is self-defended.

2.7 Summary

This chapter discusses some theoretical models based on abstract immunological processes. Among these processes, the clonal selection theory has been studied since the 1950s to understand how certain types of B and T lymphocytes are selected for destruction of specific antigens invading our body.

The IN theory has been investigated extensively by many theoretical immunologists, and three generations of this theory were developed to simulate complex dynamic behavior of the immune system. Another approach used multiepitope IN model and mapped into a PDP as a neural network. The GC is a complex system and hence, it is difficult to model all its internal mechanisms. The more the functions of the GC become apparent, the more is the scope of understanding its immunological behavior. Matzinger proposed DT so as to explain how the immune system responds although APC-activating alarm signals arise from the damaged tissues.

This chapter ends with highlighting a number of features observed in adaptive immunity from the computational point of view. Thus, the purpose of this chapter is to provide different models so that the characteristics of immune response can be better understood.

2.8 Review Questions

1. Explain the clonal expansion process. What is the purpose of clonal expansion? How does clonal selection take place?
2. What will happen if there is no mutation in the immune process?
3. What is an idiotypic network?
4. What is the main idea behind IN theory?
5. Discuss the following statement "there is no intrinsic difference between an antigen and an antibody in an idiotypic network."
6. What motivated the development of different generations of IN models?
7. Explain the main features of the different IN generations.
8. Associate each one of the following characteristics to one of the three IN generations discussed in this chapter.
    a. Try to predict the amount of each kind of antibodies (clone) present in the blood.
    b. Try to explain and describe specific clonal immune responses to external antigens.
    c. This theory competes with clonal selection theory.
    d. This theory is based on the postulate that interactions of B cells with the network are determined by their affinity to preexisting soluble antibodies and self-antigens.
    e. Interactions of B cells with the network have distinct functional consequences according to a bell-shaped dose response curve.
    f. Try to model autonomous behavior of the immune system.
    g. It considers free antibodies as well as active B and T cells and self-antigens.
    h. These models involved the notion of network dynamics and concept of metadynamics.
    i. It brought two views of the immune system into compatibility, clonal selection and network theories.
    j. Introduced the concept of a central immune system.
    k. It incorporates B and T cells cooperation.
    l. B cell activation process is considered to occur in two steps: induction and cooperation.

9. Illustrate with a diagram, a multiepitope IN approach as a PDP with input or output and hidden layers.
10. Mention two different models of the GC? Explain how they differ.
11. Why is somatic hypermutation necessary during the B cell affinity maturation process?
12. Explain the main ideas of DT and the integrity model.
13. Explain the four different stimuli categories proposed in DT.
14. Mention the computational features of adaptive immune systems.
15. Try to formulate some computational aspects of the immune system different from the ones explained in this chapter.

References

Berek, C., A. Berger and M. Apel. Maturation of immune response in germinal centers. *Cell*, 67, 1121–1129, 1991.

Bersini, H. Self-asssertion vs self-recognition: A tribute to Francisco Varela. *Proceedings of the first ICARIS conference,* Canterbury, pp. 107–112, 2002.

Bretscher, P. and M. Cohn. A theory of self-nonself discrimination. *Science,* 169(3950), 1042–1049, 1970.

Burnet, F. M. *The Clonal Selection Theory of Immunity.* Cambridge University Press, London, 1959.

Burnet, F. M. The evolution of receptors and recognition in the immune system. In P. Cuatrecasas and M. F. Greaves (Eds.), *Receptors and Recognition*, Vol. 1, Ser. A, pp. 33–58, Chapman & Hall, London, 1976.

Camacho, S. A., M. H. kosco-Vilbois and C. Berek. The dynamic structure of the germinal center. *Immunol. Today*, 19, 511–514, 1998.

Casamayor-Palleja, M., A. Gulbranson-Judge and I. C. M. MacLennan. T cells in the selection of germinal center B cells. In M. Ferrarini and F. Caligaris-Cappio (Eds.), *Human B Cell Populations.* Chemical Immunology. Karger, Basel, Vol. 67, pp. 27–44, 1997.

Celada, F. and P. E. Seiden. Affinity maturation and hypermutation in a simulation of the humoral immune response. *Eur. J. Immunol.,* 26, 1350–1358, 1996.

Coutinho, A. Beyond clonal selection and network. *Immunol Rev.,* 110 (Aug), 63–87, 1989.

Dasgupta, D. (Ed.) *Artificial Immune Systems and Their Applications.* Springer, Berlin, 1999.

De Boer, R. J. *Clonal Section Versus Idiotypic Network Models of the Immune System: A Bioinformatics Approach.* PhD thesis, University of Utrecht, the Netherlands, 1989.

Dembic, Z. Immune system protects integrity of tissues. *Mol. Immunol.,* 37(563), 8, 2000.

Hightower, R. *Computational Aspect of Antibody Gene Families.* Dissertation, Technical Report, The University of New Mexico, Albuquerque, NM, 1996.

Hoffmann, G. W. A theory of regulation and self-nonself discrimination in an immune network. *Eur. J. Immunol.,* 5(9), 638, 1975.

Jacob, J., G. Kelsoe, K. Rajesky and U. Weiss. Intraclonal generation of antibody mutants in germinal centers. *Nature,* 354, 389–392, 1991.

Jerne, N. K. Clonal section in a lymphocyte network. In G. M. Edelman (Ed.), *Cellular Selection and Regulation in the Immune System*, Raven Press, New York, pp. 39–48, 1974.

Kepler, T. B. and A. S. Perelson. Cyclic re-entry of germinal center B cells and the efficiency of affinity maturation. *J. Theor. Biol.*, 164, 37–64, 1993.

Kesmir, C. and R. J. De Boer. A mathematical model on GC kinetics and termination. *J. Immunol.*, 163, 2463–2469, 1999.

Kuby, J., R. A. Goldsby, T. J. Kindt and B. A. Osborne. *Immunology*. Fourth Edition, W. H. Freeman, San Francisco, CA, 2000.

Leanderson, T., E. Kallberg and D. Garry. Expansion, selection and mutation of antigen-specific B cells in germinal centers. *Immunol. Rev.*, 126, 47–61, 1986.

Liu, Y. J., D. E. Joshua, G. T. Williams, C. A. Smith, J. Gordon and I. C. M. MacLennan. Mechanism of antigen-driven selection in germinal center. *Nature*, 342, 929–31, 1989.

MacLennan, I. C. M. Germinal centers. *Annu. Rev. Immunol.*, 12, 117–139, 1994.

Matzinger, P. Tolerance, danger, and the extended family. *Ann. Rev. Immunol.*, 12, 991–1045, 1994.

Matzinger, P. The danger model: A renewed sense of self. *Science*, 296, 301–305, 2002.

Neumann, A. U. *Dynamical Transitions and Percolation in Network models of the Immune Response*. Phd thesis, Bar-Ilan University, Isreal, 1992.

Oprea, M. and A. S. Perelson. Somatic mutation leads to efficient affinity maturation when centrocytes recycle back to centroblasts. *J. Immunol.*, 158, 5155–5162, 1997.

Pramanik, S., R. Kozma and D. Dasgupta. Neural representation of germinal center dynamics and its application in data clustering. *Proceedings of the International Joint Conference on Neural Networks*, Honolulu, May 12–17, 2002.

Stewart, J. and J. Carneiro. The central and the peripheral immune systems: What is the relationship? In D. Dasgupta (Ed.), *Artificial Immune Systems and Their Applications*. Springer, Berlin, pp. 47–64, 1999.

Stewart, J., F. J. Varela and A. Coutinho. The relationship between connectivity and tolerance as revealed by computer simulation of the immune network: Some lessons for an understanding of autoimmunity. *J Autoimmun.*, (Jun, 2 Suppl), 15–23, 1989.

Varela, F., A. Coutinho, B. Dupire and N. Vaz. Cognitive networks: Immune and neural and otherwise. In A. Perelson (Ed.), *Theoretical Immunology: Part Two, SFI Studies in the Science of Complexity*, Addison-Wesley, Vol. 2, 359–371, 1988.

Varela, F. and A. Coutinho. Second generation immune networks. *Immunol. Today*, 12(5), 159–166, 1991.

Varela, F. J. and J. Stewart. Dynamics of a class of immune networks. I. Global stability of idiotype interactions. *J Theor Biol.*, 144(1), 93–101, 1990.

Vertosick, F. T. and R. H. Kelly. Immune network theory: A role for parallel distributed processing? *Immunology*, 66, 1–7, 1989.

Vertosick, F. T. and R. H. Kelly. The immune system as a neural network: A multi-epitope approach. *J. Theor. Biol.*, 150, 225–237, 1991.

Von Boehmer, H. and P. Kisielow. Self-nonself discrimination by T cells. *Science*, 248(4961), 1369–1373, 1990.

Wiesbuch, G., R. J. De Boer and A. S. Perelson. Localized memories in idiotypic network. *J. Theor. Biol.*, 146, 483, 1990.

Chapter 3

Immunity-Based Computational Models

The field of immunological computation (IC) or artificial immune system (AIS) has been evolving steadily (Dasgupta, 1999; Forrest et al., 1994; Tarakanov and Dasgupta, 2000) since 1985. There has been an increasing interest in the development of computational models inspired by several immunological principles (Perelson and Oster, 1979; Percus et al., 1993). Some models intend to mimic the abstract mechanisms in the biological immune system (BIS) to better understand its natural processes and simulate its dynamic behavior in the presence of antigens or pathogens; others, however, emphasize on designing artifacts—computational algorithms, techniques using simplified concepts (sometime obsolete) of various immunological processes, and functionalities (Farmer et al., 1986; Hofmeyr and Forrest, 2000; De Castro and Von Zuben, 2000; Stepney et al., 2004). Table 3.1 summarizes the mostly studied computational models of BIS, whereas the details of these are described in Chapters 4 through 6. It shows the use of specific immunological concepts in different models and their intended applications (discussed in Chapter 7).

Common terminologies that are used in most immune algorithms and their corresponding terms used in machine learning are listed in Table 3.2.

This chapter focuses on describing some common features that are used in most immunity-based models. They use computational features like shape–space representation, affinity measures, and immunity-based processes (Figure 3.1).

Table 3.1   Immunity-Based Computational Models and Specific Immunological Concepts

Immunological Concepts and Entities	Immunity-Based Models	Computational Problem
Self or nonself recognition T cell	Negative selection algorithms (Forrest et al., 1994)	Anomaly, fault, and change detection
Idiotypic network, immune memory, and B cell	Immune network theory (Hunt and Cooke, 1995)	Learning (supervised and unsupervised)
Clonal expansion, affinity maturation, and B cell	Clonal selection algorithm (De Castro and Von Zuben, 2000)	Search and optimization
Innate immunity	DT (Aickelin and Cayzer, 2002)	Defense strategy

Table 3.2   Machine Learning versus Immunology Terminology

Machine Learning	Immune Models
Detectors, clusters, classifiers, and strings	T cells, B cells, and antibodies
Positive samples, training data, and patterns	Self-cells, self-molecules, and immune cells
Incoming data, verifying data samples, and test data	Antigens, pathogens, and epitopes
Distance and similarity measures	Affinity measure in the shape–space
String-matching rule	Complementary rule and other rules

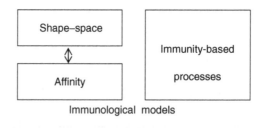

Immunological models

Figure 3.1   Major components of immunity-based models.

3.1   Shape–Space and Affinity

Perelson and Oster (1979) introduced a shape–space (or representation space) concept to represent antibody or antigen binding (see Figure 3.2). Accordingly, antigens and antibodies are characterized by their physicochemical binding properties, which are represented as coordinate points in such space, typically, a Euclidean

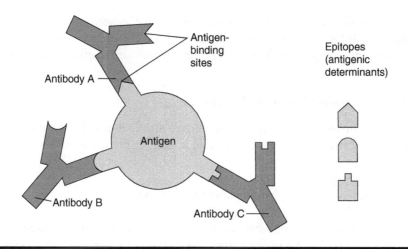

Figure 3.2   Antibody and antigen binding. An antigen may bind to several antibodies.

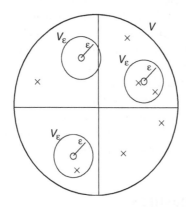

Figure 3.3   Antigens and antibodies are represented as points in an *N*-dimensional (Euclidean) space.

space (Figure 3.3). Binding properties include geometric shape, hydrophobicity, charge, etc. (Noest et al., 1997). In computational models, the notion of affinity between antibodies and antigens is defined based on a distance measure between points in the shape–space. Specifically, a small distance between an antibody and an antigen represents high affinity between them. It should be noticed that in some cases, coordinates are not given explicitly but the distance between antibodies and antigens is provided.

In Figure 3.3, the big outer circle V, crosses (X), and small inner circles V_ε represent the shape–space, antigens, and affinity (coverage) of antibodies, respectively.

Thus, ε specifies a recognition threshold; if the affinity between an antibody and an antigen (X) is less than ε (i.e., the antigen lies inside the affinity region of an antibody), then the antigen is said to match (bind) the antibody (Balthrop et al., 2002).

3.1.1 Representation Schemes

The entities involved in immune algorithms are mainly B- and T cells, antibodies, and antigens. Representations that are used in most immune algorithms are

- Binary strings
- Strings over finite alphabets (other than binary)
- Real-valued vectors
- Hybrid representation where each entity consists of several features and each feature may be of a different type; for instance, integer, real value, boolean value, or categorical information

The binary representation, in general, can subsume other representations, that is, any data type can be represented as a sequence of bits in the memory of a computer (although their treatment differs). In theory, any matching rule defined on a high-level representation can be expressed as a binary matching rule. Many models use binary representation. Although binary string representation has some advantages (any type of data can be represented in binary form, it is easy to analyze, and it is good to represent categorical data), it also has limitations—it is difficult to interpret in the original problem space, presents scalability issues, and is difficult to directly apply on some conventional techniques that assume continuous spaces (Gonzalez et al., 2003). Therefore, the other types of representations have been investigated for use in immune algorithms (Stibor et al., 2005).

3.2 Affinity Measures

To define the notion of affinity between a T or B cell and an antigen, different similarity or distance measures are introduced. If a string representation is used, a Hamming distance may be suitable. However, in the case of binary strings, other distance measures have been used. Although a Euclidean distance may be used when using a real-valued vector representation, other distance measures have also been used.

A T cell is considered to detect foreign antigens in a certain region of the shape–space. This is due to the fact that antigen matching is not exact, but approximate. Thus, a T cell will match variations of a specific antigen. A T cell model is called a detector; a detector is characterized by a set of attributes and a "matching rule," which is based on a distance measure (Gonzalez et al., 2003). Generally, a detector can be implemented as either a production rule or a neural network or a software

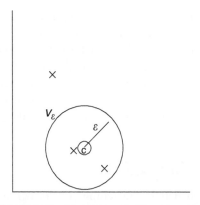

Figure 3.4 **A detector represents a subset of the shape–space, in fact, a subset of the nonself space. A detector corresponds to an *N*-dimensional hypersphere; a hypersphere may be represented by its center (c) and radius (ε), where V_ε denotes the hypersphere's volume.**

agent, among others; however, such type of implementations may be difficult to analyze (Hofmeyr, 1999).

Usually, a detector describes a region of the shape–space; therefore, it may be defined by a formal representation of a subspace of the shape–space. For instance, when the shape–space is a Euclidean space, a detector could have the shape of a hypersphere; thus, it may be defined by a hypersphere's equation, which is determined by two parameters—its center and radius (Figure 3.4). Another example of a detector is a hyperrectangle, which can be characterized by two of its (opposite) corners, similar to the way a rectangle is specified in two dimensions by the left-hand side lower corner and right-hand side upper corner. Thereby, in the case of binary representations, a detector may be defined by a binary string (which may be thought of as the "center" of the detector) and threshold value; the detector will represent all the strings that are at a distance from the center below such threshold.

In specific applications, a set of variables will define the shape–space. Therefore, a detector is defined on a data space; and although a detector could be defined as one of the points in the dataset, in general, this is not the case. A detector is rather defined as a set of data points. Intuitively, the "size" of the set associated to a detector will indicate its specificity; thus, the larger the set, the more general the detector will be. Accordingly, the smaller the set of points covered by a detector, the more specific the detector.

A matching rule is a key concept in immune algorithms, because it is used to determine when a detector matches a data item. A particular immune algorithm is distinguished by the way entities are modeled, as well as the matching rule and the mechanisms to generate the detectors. It is important to note that the matching rule is used in both detector generation and detection.

Regardless of the representation, a matching rule M may be defined in terms of an affinity measure. Accordingly, $d\,M\,x$ that denotes "d matches x" is defined as $d\,M\,x$ if and only if x "belongs" to the set defined by the detector d.

The quotations mean that it is necessary to define a notion of a point belonging to a set. In classical set theory, a point either belongs to a set or does not; however, a different notion like the one in fuzzy set theory may be used, where a degree of membership to the universal set is defined (Gonzalez, 2003).

3.3 String-Matching Rules

A matching rule, which defines "matching" or "recognition," and the distance measure that the former is based on are the cornerstones in any detection, classification, or recognition algorithms. The choice of a matching rule depends on the representation scheme and type of data. For instance, if you are dealing with categorical data, then a string representation may be more suitable, and a matching rule such as r-contiguous bits (*rcb*) or r-chunks can be used (Percus et al., 1993). In this subsection, several string-matching rules are described in detail.

3.3.1 Hamming Distance

The Hamming distance between two strings is defined as the number of different characters between the two strings. The Hamming distance $h(x,y)$ between two strings x and y is expressed as

$$h(x, y) = \sum_{i=1}^{N} \left(\overline{X_i \oplus Y_i} \right)$$

where N is the string length, X_i and Y_i denote the ith bit of string n and y, respectively, $X_i \oplus Y_i$ the *Xor* logic operation, when dealing with binary strings; for other alphabets, $X_i \oplus Y_i$ is zero if the two symbols are equal and one otherwise.

3.3.2 Binary Distance

Some extensions of the Hamming distance, for binary strings, have been proposed based on the relative number of bits that match or differ; such extensions are based on the following basic measures:

$$a = \sum_{i=1}^{N} \zeta_i, \quad \zeta_i = \begin{cases} 1, & \text{if } X_i = Y_i = 1 \\ 0, & \text{otherwise} \end{cases}$$

$$b = \sum_{i=1}^{N} \zeta_i, \quad \zeta_i = \begin{cases} 1, & \text{if } X_i = 1, \ Y_i = 0 \\ 0, & \text{otherwise} \end{cases}$$

$$c = \sum_{i=1}^{N} \zeta_i, \quad \zeta_i = \begin{cases} 1, & \text{if } X_i = 0, \ Y_i = 1 \\ 0, & \text{otherwise} \end{cases}$$

$$d = \sum_{i=1}^{N} \zeta_i, \quad \zeta_i = \begin{cases} 1, & \text{if } X_i = Y_i = 0 \\ 0, & \text{otherwise} \end{cases}$$

X_i and Y_i denote the ith bit of string x and y, respectively; a counts the number of 1s that match at the same positions of both strings; similarly, d counts the number of 0s that match at the same positions of both strings; b counts the number of 1s in string x that do not match string y; and c counts the number of 0s in string x that do not match string y. These values are combined to define different similarity functions as follows:

1. Russel and Rao

$$f = \frac{a}{a + b + c + d}$$

2. Jacard and Needham

$$f = \frac{a}{a + b + c}$$

3. Kulzinski

$$f = \frac{a}{b + c + 1}$$

4. Sokal and Michener

$$f = \frac{a + d}{a + b + c + d}$$

5. Rogers and Tanimoto

$$f = \frac{a+d}{a+d+2(b+c)}$$

6. Yule

$$f = \frac{ad-bc}{ad+bc}$$

3.3.3 Edit Distance

The edit distance between two strings $s1$ and $s2$ is defined as the minimum number of string transformations required to change $s1$ into $s2$ where the possible string transformations are (i) changing a character, (ii) inserting a character, and (iii) deleting a character. The edit distance is also called "Levenshtein" distance, and it is a generalization of the Hamming distance.

3.3.4 Value Difference Metric

Value difference metric (VDM) distance is defined as (Hamaker and Boggess, 2004)

$$\text{VDM}(x, y) = \sum_{i=1}^{N} \text{vdm}(x_i, y_i) \cdot \text{weight}(x_i)$$

where

$$\text{vdm}(x_i, y_i) = \sum_{c \in C} \left(P(c \mid x_i) - P(c \mid y_i) \right)^2$$

and

$$\text{weight}(x_i) = \sqrt{\sum_{c \in C} P(c \mid x_i)^2}$$

$P(c \mid x_i)$ denotes the probability that x_i be equal to the character c in the alphabet C.

3.3.5 Landscape-Affinity Matching

This matching rule was proposed to capture the ideas of matching biochemical and physical structures and approximate matching in the immune system (Harmer et al., 2002). An input string and antibody strings are sampled as bytes and converted into positive integer values to generate a landscape. Two strings are then compared using a sliding window. In fact, three different similarity measures are defined as follows:

1. Difference-matching rule

$$f_{\text{difference}} = \sum_{i=1}^{N} |X_i - Y_i|$$

2. Slope-matching rule

$$f_{\text{slope}} = \sum_{i=1}^{N} |(X_{i+1} - X_i) - (Y_{i+1} - Y_i)|$$

3. Physical matching

$$f_{\text{physical}} = \sum_{i=1}^{N} (X_{i+1} - Y_i) + 3 \times |\mu| \min(\forall i, (X_i - Y_i))$$

3.3.6 R-Contiguous Bits Matching

The *rcb* matching rule, introduced by Percus et al. (1993), is defined as follows: If x and y are equal-length strings defined over a finite alphabet, *match*(x, y) is true if x and y agree in at least r contiguous locations.

As an example, if x = ABA*DCB*AB and y = CAG*DCB*BA, then *match*(x, y) is true for $r \leq 3$ and false for $r > 3$. In the case of binary strings, a matching rule typically used is *rcb* (Forrest et al., 1994) where a detector d is specified by a binary string c and threshold value r, and d matches a string x if *rcb* of c matches the corresponding bits (at the same positions) of x. It was originally designed to consider approximate matching between two strings. The choice of *rcb* simplifies mathematical analysis and is a good model for approximate T cell matching. The parameter r determines a detector's degree of specificity; the smaller the value of r, the more general is the detector.

3.3.7 R-Chunk Matching Rule

Balthrop et al. (2002) introduced a generalization of *rcb* matching rule called *r*-chunk matching rule. As in *rcb* matching rule, a detector is specified by a binary string *c* and parameter *r*. An *r*-chunk detector *d* is said to match a string *x* if all bits of *c* are equal to the *r* bits of *x* in the window specified by *c*. In contrast to the *rcb* matching rule, the *r*-chunk approach allows the use of detectors of any size. This fact has an improvement on the ability of the detectors to cover the self-space. The difference from *rcb* rule is that the matching window is specified for each individual detector. A group of *r*-chunk detectors that cover all possible windows has the same effect as an *rcb* detector.

In mathematical terms, $x = e_1 e_2 \ldots e_m$ (an element in the shape–space) and $d = (p; d_1 d_2 \ldots d_r)$ with $r \le m$, $p \le m - r + 1$ match according to the *r*-chunk rule if and only if $e_i = d_i$ for $i = p, \ldots, p + r - 1$. In other words, element *x* matches detector *d* if, at position *p*, there is a sequence of length *r* where all the characters are identical.

3.3.8 Real-Valued Vector Matching Rules

Some distance measures that have been used to define matching rules in real-valued vector representation are explained in the following sections.

3.3.8.1 Euclidean Distance

A Euclidean distance is defined as

$$d(x, y) = \sqrt{\sum_i (x_i - y_i)^2} = \|x - y\|$$

Euclidean distance can be modified when all the dimensions do not have equal weights by multiplying each component of the vectors by specific weights. Other distance measures can be used to define real-valued matching rule in a similar way to Euclidean distance. The choice of distance measures mainly relies on the type of data and domain knowledge of the specific application. Several distance measures summarized by Hamaker and Boggess (2004) are presented in the following sections.

3.3.8.2 Partial (Euclidean) Distance

It is a variation of the Euclidean distance, and it is defined over some elements of a vector, as opposed to the whole vector. It is equivalent to the Euclidean distance projected on a lower dimensional subspace of the original space. In other words, the Euclidean distance is not calculated over all the elements of the vector, but it is calculated taking into account only some elements. This is similar to the way some of all the available bits are used in matching two strings, in the *rcb* rule. Here, also the size of a window needs to be specified. In an *N*-dimensional space, the partial distance between *x* and *y*, for a window size *w*, is defined as

$$d(x, y) = \sum_{i=1}^{w}(x_{s_i} - y_{s_i})^2$$

where s_i belongs to {1, 2, ..., *N*}.

3.3.8.3 Minkowski Distance

It is also known as λ-norm distance, and it is defined as

$$d(x, y) = \left(\sum |x_i - y_i|^\lambda\right)^{\frac{1}{\lambda}}$$

When $\lambda = 1$, it becomes the Manhattan distance, also known as the city block distance. If $\lambda = 2$, it is equivalent to a Euclidean distance.

3.3.8.4 Chebyshev Distance

It is also known as the infinity norm distance, denoted by D_∞, and it is defined as the maximum of the differences for all features

$$d(x, y) = \max\{|x_i - y_i| \quad \text{for } i = 1, ..., n\}$$

3.3.9 Mixed Representation

Some distance measure defined for mixed data, that is, continuous and categorical data, are explained in the following sections.

3.3.9.1 Heterogeneous Euclidean-Overlap Metric

Heterogeneous Euclidean-overlap metric (HEOM) distance is defined for mixed data, that is, continuous and categorical data.

$$\text{HEOM}(x, y) = \sqrt{\sum_{i=1}^{N} \text{heom}(x_i - y_i)^2}$$

where

$$
\text{heom}(x_i, y_i) = \begin{cases} \text{overlap}(x_i, y_i), & \text{if the } i\text{th variable is categorical} \\ \dfrac{|x_i - y_i|}{\text{range}_i}, & \text{if } i\text{th variable is real} \end{cases}
$$

overlap(*x, y*) denotes the Hamming distance defined as

$$
\text{overlap}(x_i, y_i) = \begin{cases} 0, & x_i = y_i \\ 1, & x_i \neq y_i \end{cases}
$$

and range$_i$ is a scaling factor for the *i*th continuous variable.

3.3.9.2 Heterogeneous Value Difference Metric

Heterogeneous value difference metric (HVDM) metric is defined as

$$
\text{HVDM}(x, y) = \sqrt{\sum_{i=1}^{N} \text{hvdm}(x_i - y_i)^2}
$$

where

$$
\text{hvdm}(x_i, y_i) = \begin{cases} \sqrt{\text{vdm}(x_i, y_i)}, & i\text{th variable is categorical} \\ \dfrac{|x_i - y_i|}{\text{range}_i}, & i\text{th variable is real} \end{cases}
$$

3.3.10 Considerations about Representation

In Gray encoding, two consecutive strings differ only by 1 bit; thereby, affinity in problem space is to some extent maintained in shape–space. A real number *x* in [0,1] can be represented in a binary encoding, using the transformation *floor*(255*x* + 0.5) and then encoding it in 8 bits. Therefore, binary encoding is not suitable to achieve good generalization, because matching rules should accurately represent data proximity in the problem space.

Matching rules also have effects in searching the shape–space. For instance, *rcb* matching rule produced a gridlike shape; *r*-chunk matching rule generated similar but simpler shapes; Hamming distance and Rogers and Tanimoto (R&T) matching rules produced a "fractal"-like shape. The shape of areas covered by *rcb* and *r*-chunk matching rules were not affected by changing encoding from binary to Gray. This was not really unexpected because similarity between two real values is not reflected in their binary representations (Gonzalez et al., 2003; Ji and Dasgupta, 2004).

3.4 Affinity Maturation

Affinity maturation is a process of variation and selection that occurs among stimulated B cells. Variation is achieved by somatic hypermutation, whereas the selection among cloned B cells is performed to better match the pathogen at hand (Celada and Seiden, 1996).

Generally, when two immune entities match, for instance, an antibody matches an antigen or two B cells stimulate one another, they undergo some proliferation and differentiation processes, which correspond to an abstraction of clonal selection. Also, somatic hypermutation is used as a mechanism to introduce variation in a population of immune entities. Depending on their representation, such a process consists of applying some variation process, analogous to a mutation operator in an evolutionary algorithm. For example, if immune entities are represented as binary strings, a bit mutation operator can be applied. In contrast, if a real-value representation is considered, a Gaussian mutation may be suitable. Generally, any type of variation operator can be applied as long as it guarantees a good exploration of the shape–space.

3.5 Solving Problems Applying Immunity-Based Models

To apply an immunity-based model to solve problems in a specific domain, one should select a specific model according to the type of problem that is being solved (De Castro and Timmis, 2002; Harmer et al., 2002). The first step is to identify the salient features of the problem and decide how they can be modeled; and a representation scheme for each entity should be chosen, specifically, a string-, real-valued vector, or hybrid representation. Next, an appropriate affinity measure should be used according to the defined matching rule and immune algorithm. In many immune algorithms, it is necessary to introduce new entities to accomplish the desired search and optimization task. Figure 3.5 shows these steps to solve a problem by using an immunity-based model.

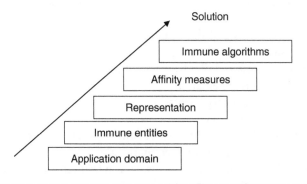

Figure 3.5 Problem solving by using immunity-based models.

3.6 Summary

This chapter describes the general aspects of immunity-based models. First, the shape–space concept, which is very important to represent the elements that are involved in immune processes, is introduced. Then, some matching rules that are very important to determine the affinity or similarity between immune elements are explained in detail. Accordingly, some distance measures that have been used in immune modeling are presented. Finally, some general operations that are inspired in immunology and have been applied in artificial immune models are summarized.

3.7 Review Questions

1. List the main elements necessary to define an immune model. Explain the relations among such elements.
2. Consider a particular representation for the immune elements. Then consider a particular immune algorithm and change the immune representation. Does the performance of the algorithm change? In general, do the results depend on a particular representation?
3. Define other matching rules different from the ones described in this chapter.
4. Consider a particular representation for the immune elements (same as question 2).
5. Define three different types of clonal selection operators.
6. Define three different types of somatic hypermutation.
7. Define some ways to introduce new random elements in the population.
8. Define three different types of affinity maturation.
9. Define other representations for immune entities.
10. Define other formal representations for the immune entities. What advantages do your representations have with respect to those that have been used in immune algorithms?
11. List the advantages and disadvantages of processing strings with arbitrary length versus fixed length strings.
12. List the advantages and disadvantages of using
    a. Binary string representation
    b. Real-valued vector representation
    c. Hybrid representation
13. Why would a Hamming distance be suitable when a string representation is used? Is it suitable for other representations?
14. Is it possible to define a detector using a hybrid representation? Illustrate a hybrid representation by an example.
15. Associate immune elements to each arrow in the following figure.

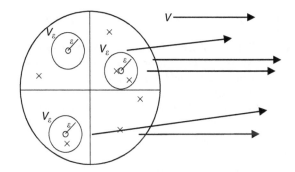

16. Complete the following statements:
    a. The _____ the set points a detector covers, the more _____ the detector is.
    b. In affinity maturation, the _____ the affinity of a B cell for pathogens present, the _____ likely it is that the B cell will clone.
    c. Any type of variation operator can be applied as long as it guarantees a good _____ of the _____.
17. List the three main features that characterize a particular immune algorithm.
18. Why may the choice of a matching rule depend on a particular type of data? Could you give an example where a specific matching rule does not work on a particular data?
19. Give a formal representation where the best string-matching rule could be
    a. Edit distance
    b. Hamming distance
    c. Binary distance
    d. Value difference metric
    e. Landscape-affinity matching
    f. *rcb* matching
    g. R-chunk matching rule
20. Define and solve a problem by using immune models. Write the necessary information to specify your problem with respect to the elements in the boxes in Figure 3.5.

References

Aickelin, U. and S. Cayzer. The danger theory and its application to artificial immune systems. *Proceedings of 1st International Conference on Artificial Immune Systems (ICARIS)*. University of Kent at Canterbury, U.K., September 9–11, 2002.

Balthrop, J., F. Esponda, S. Forrest and M. Glickman. Coverage and generalization in an artificial immune system. *Proceedings of the Genetic and Evolutionary Computation Conference (GECCO)*, Morgan Kaufmann, New York, pp. 3–10, 2002.

Celada, F. and P. E. Seiden. Affinity maturation and hypermutation in a simulation of the humoral immune response. *Eur. J. Immunol.*, 26, 1350–1358, 1996.

Dasgupta, D. (Ed.). *Artificial Immune System and Their Applications*. Springer, Berlin, pp. 3–23, 1999.

De Castro, L. N. and F. J. Von Zuben. The clonal selection algorithm with engineering applications. *Proceedings of GECCO*, Las Vegas, pp. 36–39, 2000.

De Castro, L. N. and J. Timmis. *Artificial Immune Systems: A New Computational Intelligence Approach*. Springer, Heidelberg, 2002.

Farmer, J. D., N. H. Packard and A. S. Perelson. The immune system, adaptation, and machine learning. *Physica D*, 22, 187–204, 1986.

Forrest, S., A. S. Perelson, L. Allen, and R. Cherukuri. Self-nonself discrimination in a computer. *Proceedings of the 1994 IEEE Symposium on Research in Security and Privacy*, IEEE Computer Society Press, Los Alamitos, CA, 1994.

Gonzalez, F., D. Dasgupta and J. Gomez. The effect of binary matching rules in negative selection. *Proceedings of the Genetic and Evolutionary Computation Conference (GECCO)*, Lecture Notes in Computer Science, Springer, Berlin, Vol. 2723, pp. 198–209, July 2003.

Gonzalez, F. *A Study of Artificial Immune Systems Applied to Anomaly Detection*. PhD thesis, The University of Memphis, Memphis, TN, 2003.

Hamaker, J. S. and L. Boggess. Non-euclidean distance measures in AIRS, an artificial immune classification system. *Proceedings of 2004 Congress on Evolutionary Computation (CEC 2004)*, Portland, OR, pp. 1067–1073, June 2004.

Harmer, P., G. Williams, P. D. Gnush and G. Lamont. An artificial immune system architecture for computer security applications. *IEEE Transaction on Evolutionary Computation*, 6(3), 252–280, 2002.

Hofmeyr, S. A. *An Immunological Model of Distributed Detection and its Application to Computer Security*. PhD thesis, University of New Mexico, Albuquerque, NM, 1999.

Hofmeyr, S. A. and S. Forrest. Architecture for an artificial immune system. *Evol. Comput.*, 8(4), 443–473, 2000.

Hunt, J. E. and D. E. Cooke. An adaptive and distributed learning system based on the immune system. *Proceedings of the IEEE International Conference on Systems Man and Cybernetics*, Las Vegas, pp. 2494–2499, 1995.

Ji, Z. and D. Dasgupta. Real-valued negative selection algorithm with variable-sized detectors. *Lecture Notes in Computer Science 3102, Proceedings of GECCO*, pp. 287–298, 2004.

Noest, A. J., K. Takumi and R. de Boer. Pattern formation in B-cell immune network: Domains and dots in shape-space. *Physica D*, 105, 285–306, 1997.

Percus, J. K., O. E. Percus and A. S. Perelson. Predicting the size of the T-cell receptor and antibody combining region from consideration of efficient self-nonself discrimination. *Proceedings of National Academy of Sciences USA 90*, Las Vegas, pp. 1691–1695. 1993.

Perelson, A. S. and G. F. Oster. Theoretical studies of clonal selection: Minimal antibody repertoire size and reliability of self- non-self discrimination. *J. Theor. Biol.*, 81, 645–670, 1979.

Stepney, S., R. E. Smith, J. Timmis and M. Tyrrell. Towards a conceptual framework for artificial immune systems. *Proceedings of Third International Conference on Artificial Immune Systems (ICARIS)*, pp. 53–64, 2004.

Stibor, T., J. Timmis and C. Eckert. A comparative study of real-valued negative selection to statistical anomaly detection techniques. *ICARIS*, pp. 262–275, 2005.

Tarakanov, A. and D. Dasgupta. A formal model of an artificial immune system. *BioSystems*, 55, 151–158, 2000.

Chapter 4

T Cell–Inspired Algorithms

This chapter presents change/anomaly detection techniques inspired by the T cell censoring and maturation process in the biological immune system. First, the main process of negative selection (NS) that results in self/nonself discrimination is described. Then, the important features of artificial NS are presented. Finally, different variations of negative selection algorithm (NSA) have been considered. There is a significant amount of work on this topic area; therefore, for better readability, this chapter has been divided into three parts.

4.1 Self/Nonself Discrimination

An important mechanism of the adaptive immune system is the "self/nonself recognition" (Coutinho, 1980). The immune system is able to recognize which cells are its own (self) and which are foreign (nonself); thus, it is able to build its defense against the attacker instead of self-destructing. As mentioned earlier (Chapter 1), T cells of enormous diversity are first assembled with a "pseudorandom genetic rearrangement process" and those that recognize self-cells are eliminated before the rest are deployed into the immune system to recognize and attack foreign pathogens. Therefore, T cells go through a process of selection that ensures that they are able to recognize nonself peptides presented by major histocompatibility complex (MHC). This process has two main phases: positive selection (PS) and NS.

During the PS phase, T cells are tested for recognition of MHC molecules expressed on the cortical epithelial cells. If a T cell fails to recognize any of the MHC molecules, it is discarded; otherwise, it is kept.

The purpose of NS is to test for tolerance of self-cells. T cells that recognize the combination of MHC and self-peptides fail this test. This process can be viewed as a filtering of a big diversity of T cells; only those T cells that do not recognize self-peptides are kept (Kappler et al., 1987).

4.2 Negative Selection Algorithms

Forrest et al. (1994) proposed a computational model of self/nonself discrimination, which is called the "NSA or NS algorithm." This algorithm models the T cell maturation process that occurs in the thymus. Several variations of NSAs have been proposed after the original version was introduced (Forrest et al., 1994); however, the main features of the original algorithm still remain. Particularly, the goal of NS is to cover the nonself space with an appropriate set of detectors (shown in Figure 4.1).

Two important aspects of an NSA are as follows:

1. The target concept of the algorithm is the complement of a self-set.
2. The goal is to discriminate between self and nonself patterns, while only self-samples are available (one-class learning; Tax, 2001).

There are two steps in NSAs as follows: "detector generation" and "nonself detection." In the first step, a set of detectors is generated by some random-ized process that uses a collection of self as the input. Candidate detectors that match any of the self-samples are eliminated, whereas unmatched ones are kept.

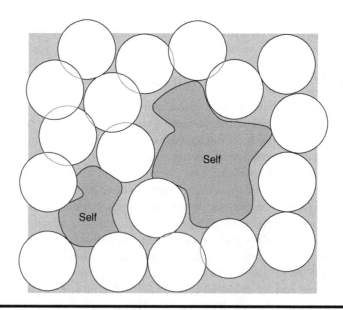

Figure 4.1   Illustration of the self and nonself regions.

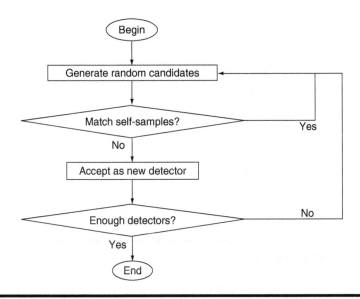

Figure 4.2 **Detector generation process—censoring phase of NSAs.**

"Matching rules" (described in Chapter 3) are usually designed inspired by T cell/ antigen affinity measures. A specific NSA is characterized by the way candidate detectors are generated and the criteria used to select the detectors. Figure 4.2 shows the major steps in an NSA.

In the detection stage, the stored detectors (generated in the first stage) are used to check whether new incoming samples correspond to self or nonself instances. If an input sample matches a detector, then it is identified as part of nonself, which in most applications, means that an anomaly/change has occurred (see Figure 4.3).

Basic NS Algorithm: Generic Negative Selection Algorithm

Input: $S \subseteq U \equiv$ Self or normal data, $l, r \in N$, where l is the string length and r is a matching threshold

Output: a set of detectors $D \subseteq U$

1 **begin**
2   Generate a set (D) of detectors, such that each fails to match any element in S.
3   Monitor new sample $\delta \in U$ by continually checking the detectors (in D) against δ. If any detector matches δ, classify it as nonself.
4 **end**

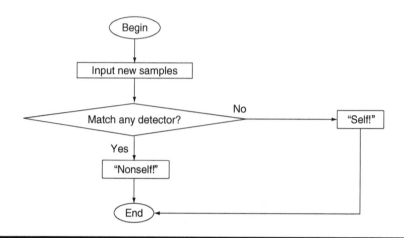

Figure 4.3   Monitoring phase of an NSA.

A particular NSA is characterized by the way detectors are represented, the rule used to determine the match between a sample and detectors, and the mechanisms to generate and discard self-reactive candidate detectors. Most works on NSAs have assumed one of the following schemes for encoding problem space: binary and real-valued vector representation.

The remainder of the sections are grouped under three parts: the first part (Sections 4.3 through 4.5) describes various detector generation schemes using string representation in NSA algorithms, the second part (Sections 4.6 through 4.8) covers detector generation schemes for real-valued and hybrid representation, and the third part (Section 4.9) covers the concepts of negative databases (NDB) and algorithms to generate NDBs.

4.3   Negative Detector Generation Schemes

In string representation, each detector is represented as a string of fixed length over a finite alphabet (for binary, 0 and 1). Different approaches have been developed to generate negative detectors having varying degree of complexity. The algorithm described first is an exhaustive approach, which appears to be analogous to the natural NS process.

4.3.1   An Exhaustive Approach (Generating Detectors Randomly)

In the original description of the NSA (Forrest et al., 1994), candidate detectors were generated randomly and then tested (censored) to see if they matched any self-string. If a match was found, then such a candidate was rejected, otherwise, it was accepted. This process was repeated until a desired number of detectors were generated. A probabilistic analysis was used to estimate the number of detectors required to provide a desired level of reliability. This straightforward approach is given in the following algorithmic form (see Algorithm 1; note that although self is considered as a set, for most applications it is a collection of data/patterns rather).

**NS Algorithm 1: Exhaustive approach in
(binary) negative detector generation**

Input: l, r, $T \in N$ where $1 \leq r \leq l$ and $S \subset U$; l = string length, r = matching threshold, T = repertoire size

Output: Set $D \subset U$ detectors generated using r-contiguous bits (*rcb*) matching rule

```
1 begin
2    D := φ
3    while |D| < T do
4       Generate randomly bit string d ∈ U
5       if d does not match any string in S then
6          D := D ∪ {d}
7 end
```

The major limitation of such random generation approach appears to be computational difficulty of generating valid detectors, which grows exponentially with the size of self. Also for many choices of l ("length") and r ("recognition threshold") and compositions of self, random generation of strings for detectors may be prohibitive. Thus, this random approach is not efficient. Subsequently, a number of detector generation approaches were proposed; these are presented in detail in the following text.

4.3.2 A Dynamic Programming Approach (Linear-Time Algorithm)

Given a collection of (equal-sized) self-strings S and a matching rule with partial "matching threshold r," called the "r-contiguous bits" (*rcb*, described in Chapter 3). This algorithm works in two phases (D'haeseleer et al., 1996) as follows:

- To count recurrences to define an enumeration of all unmatched strings (not r-matched) by S in the string space for the given self-set
- To generate a detector set (according to the repertoire size) by the counting recurrence, which is used to pick the detectors at random from the set of candidate unmatched strings

This dynamic programming approach (D'haeseleer et al., 1996) runs in linear time with the size of self. To present the algorithm, some notation should be introduced first: let s denote a binary string and $\hat{s}$ the string s stripped of its leftmost bit, $s \cdot b$, where $b \in \{0,1\}$ denotes s appended with b.

Also, "a template" of order r is a size l string consisting of $l - r$ blank symbols and r fully specified contiguous bits. $t_{i,s}$ denotes a template in which the r specified bits start at position i and are given by the r-bit substring of s. A right (left) completion of a template t is a template with all the blanks to the right (left) replaced by bits. Therefore, $(a,b]$ denotes the integer interval $(a + 1) \ldots b$.

4.3.2.1 Phase I: Solving the Counting Recurrence

Let $C_i[s]$ be the number of right completions of $t_{i,s}$ unmatched by any string in S, for $1 \le i \le (l - r + 1)$. The templates in the array $C_i[s]$ will enumerate all the possible ways two strings can match each other over rcb. $C_i[s]$ can be computed recursively, the number of unmatched right completions at i can be computed based on the number of right completions at $i + 1$ as follows: if $t_{i,s}$ is directly matched in S, $C_i[s]$ is 0, otherwise, the completions of $t_{i,s}$ consist of those with a 0 bit following the r bits of s, and those with a 1 bit instead. These are exactly the number of right completions for $\hat{s} \cdot 0$ and $\hat{s} \cdot 1$, respectively.

$$c_i[s] = \begin{cases} 0, & t_{i,s} \text{ is matched in } S \\ c_{i+1}[\hat{s} \cdot 0] + c_{i+1}[\hat{s} \cdot 1], & \text{otherwise} \end{cases}$$

Also, $C_{l-r+1}[s]$ will be 0, if the template $t_{l-r+1,s}$ is matched in S, and 1 otherwise

$$c_{l-r+1}[s] = \begin{cases} 0, & t_{l-r+1,s} \text{ is matched in } S \\ 1, & \text{otherwise} \end{cases}$$

To illustrate the preceding recurrence equations, an example is presented in the following text (taken from D'haeseleer, 1996).

Let $l = 6$ and $r = 3$. Let $s_1 = 110100$ and $s_2 = 100101$ are strings in S. Consider the patterns 110***, *101**, **010*, and ***100. These patterns are matched by s_1. Thus, $C_1[110] = C_2[101] = C_3[010] = C_4[100] = 0$. Now consider the pattern **110*. This template is matched by neither s_1 nor s_2. However, the pattern **110* does not have any unmatched right completions $C_3[110] = C_4[100] + C_4[101] = 0$. The right completions of **110* are **1100 and **1101, which are matched by s_1 and s_2, respectively.

4.3.2.2 *Phase II: Generating Strings Unmatched by S*

Note that $C_1[s]$ denotes the number of unmatched l-bit strings starting with the r-bit binary string s. The total number of strings unmatched by S is

$$T = \sum_{s \in S} C_1[s]$$

$C_1[\cdot]$ can be thought of as a partitioning of the space of unmatched strings into partitions of size $C_1[s]$ for each initial r-bit string s. Given that among all the unmatched strings starting with s, $C_2[\hat{s} \cdot 0]$ have a 0 bit next, whereas $C_2[\hat{s} \cdot 1]$ have a 1 bit next. Therefore, $C_2[\cdot]$ can be seen as a subsequent partition of the original space. Likewise, $C_3[\cdot]$ to $C_{l-r+1}[\cdot]$ will define corresponding partitions of the space. Each partition after $C_{l-r+1}[\cdot]$ will consist of one single l-bit string. Thereby, unmatched strings can be matched from 1 to T, using the lexicographic (natural) order. Thus, to generate a number N_R of unmatched strings, you can just generate N_R random numbers in $\{1, \ldots, T\}$. If $k \in \{1, \ldots, T\}$, the kth unmatched string u_k is determined as follows. Perform a binary search on $C_1[\cdot]$ to find s_1 such that

$$P_1 = \sum_{s < s_1} c_1[s] < k \le Q_1 = \sum_{s < s_1} c_1[s]$$

All unmatched strings in $(P_1, Q_1]$ have s_1 as their r-bit prefix, where u_k is the string that we are interested in the partition of unmatched strings numbered $(P_1 + 1), \ldots, Q_1$; therefore, the first r bits of u_k are given by s_1.

For each $i = 2, \ldots, (l - r + 1)$, the $(r + i - 1)th$ bit of string u_k can be established by determining the partition where k falls. To illustrate this, assume that the bit at position $(r + 1)$ needs to be determined. Let us partition the interval into intervals $I_{10} = (P_1, P_1 + C_2[\hat{s}_1 \cdot 0])$ and $I_{11} = (P_1 + C_2[\hat{s}_1 \cdot 0], Q_1)$, corresponding to the strings concatenated with either a 0 or a 1. On one hand, a bit $b_1 = 0$ is concatenated to the string if k is in I_{10}, and on the other, a bit $b_1 = 1$ is added to the string if k is in I_{11}. Subsequent intervals $(P_i, Q_i]$ are then computed according to

$$P_i = \begin{cases} P_{i-1}, & \text{if } b_{i-1} = 0 \\ P_{i-1} + c_i[\hat{s}_{i-1} \cdot 0], & \text{if } b_{i-1} = 1 \end{cases} \quad \text{and} \quad Q_i = \begin{cases} P_{i-1} + c_i[\hat{s}_{i-1} \cdot 0], & \text{if } b_{i-1} = 0 \\ Q_{i-1}, & \text{if } b_{i-1} = 1 \end{cases}$$

Thus, $s_2 = \hat{s}_1 \cdot b_1 \cdot k$ is in interval (P_2, Q_2), which can again be divided into two intervals $I_{20} = (P_2, P_2 + C_3[\hat{s}_2 \cdot 0])$ and $I_{21} = (P_2 + C_3[\hat{s}_2 \cdot 0], Q_2)$. Then, bit b_2 is defined by checking if it falls in I_{20} or I_{21}. Similar processes are followed to compute the subsequent bits.

The pseudocode of this two-phase linear time algorithm is presented as follows:

NS Algorithm 2: Dynamic programming approach

Input: $l, r, t \in N$ where $1 \le r \le l$, $t = o(r)$ and $s \in S$ and $\hat{s}$ denotes string s stripped of leftmost bit, and $S \subset U$

Output: Array $C[s]$ number of unmatched strings by any string in S, where the strings are examined from left to right

Phase I: Solving the counting recurrence

```
1 begin
2   for 1 ≤ i ≤ l − r + 1
3       while |C_i [s] | ≤ |t |do
4           if t_{i,s} matches with any bit string of S then
5               C_i [s] : = 0
6           else
7               C_i [s] : = C_{i+1} [ŝ · 0] + C_{i+1} [ŝ · 1]
8       endwhile
9   endfor
10 end
```

Phase II: Generating strings unmatched by S

```
1 begin
```

2 Total number of unmatched by S: $T = \sum_{s \in S} C_1[s]$, where $C_1[s]$ denotes the number of unmatched l-bit string

3 **If** $k \in \{1, 2, ..., T\}$ **then**

4   the kth matched string u_k is determined from steps 5 to 11

5   First interval $(P_1, Q_1]$ is calculated as

$$P_1 = \sum_{s < s_1} c_1[s] < k \le Q_1 = \sum_{s \le s_1} c_1[s]$$

6   Subsequent intervals $(P_i, Q_i]$ are calculated as from steps 7 to 10

7   **for** $i = 2, ..., (l - r + 1)$

8
$$P_i = \begin{cases} P_{i-1}, & \text{if } b_{i-1} = 0 \\ P_{i-1} + c_i[\hat{s}_{i-1} \cdot 0], & \text{if } b_{i-1} = 1 \end{cases}$$

9
$$Q_i = \begin{cases} P_{i-1} + c_i[\hat{s}_{i-1} \cdot 0], & \text{if } b_{i-1} = 0 \\ Q_{i-1}, & \text{if } b_{i-1} = 1 \end{cases}$$

```
10      endfor
11   endif
12 end
```

It is important to note that this algorithm requires storing an array C that represents all the possible ways two strings can match over *rcb*. Thereby, the algorithm runs in linear time in the sizes of the self-set and a detector set for specific values of parameters l and r. However, this algorithm requires exponential time and space in r, the matching threshold. Clearly, this is a problem while dealing with long strings, and thereby, higher values of the threshold value r.

4.3.3 A Greedy Algorithm for Detector Generation

A greedy method for generating negative detectors was suggested by D'haeseleer et al. (1996), which can provide a better detector coverage of the complement space with varying degrees of computational complexities. This algorithm tries to locate detectors (instead of selecting them at random as in the second phase of the previous approach) as far apart as possible to avoid possible overlapping. At each step, the algorithm picks a detector that matches as many unmatched nonself strings as possible.

The same array structure is used as in the previous algorithm, but to construct the array C in phase I, the strings are examined from right to left, that is, from $C_{l-r+1}[]$ to $C_1[]$. A second array C' is then constructed by scanning the strings from left to right, and computing the subsequent levels using a similar recurrence relation for C. Let $D_i[s] = C_i[s] \times C'_i[s]$ be the number of unmatched fully specified bits corresponding to "template" $t_{i,s}$, where C'_l is the number of nonmatching left completions.

If a specific template has a zero entry in D, then all strings containing that template will match some string in S. Also, for all the templates with nonzero entries in D, the corresponding strings do not match with any string in S.

NS Algorithm 3: A greedy approach

Phase I: Generate valid detector templates

In this stage, two arrays, denoted as D_S and D_R, are created. Array D_S is used to keep track of the templates that the algorithm picks from during the detector construction process. The detectors with nonzero entries in D_S will be called "valid detector templates."

Phase II: Generate strings unmatched by S

D_R indicates, for each template, the number of yet unmatched strings by previously generated detectors. To generate a new detector, those templates matching the most unmatched nonself strings are selected. The array D_R is updated each time a new detector is generated by keeping C_R and C'_R (instead of D_R) and updating them incrementally.

At the beginning, R (the detectors set) is empty, and C_R and C_R' are initialized to their maximum values: $C_{R,i}[s] = 2^{(l-r+1-i)}$ and $C_{R,i}'[s] = 2^{(i-1)}$, which correspond to $D_{R,i}[s] = 2^{(l-r)}$.

When a new detector is generated, the algorithm scans D_R looking for its largest entry. If two templates have the same largest entry, then one of the templates is picked at random. Then, the algorithm traverses D_R to the left and to the right starting at such a template, each time either a "0" or a "1" is appended to the starting template, depending on which one corresponds to the template with the highest number of yet unmatched strings by R. Then, arrays C_R and C_R' are updated to reflect that a new detector has been added to R. This is done incrementally by setting those entries in C_R and C_R' that correspond to templates that match the detector to zero, and then recalculating the appropriate values for such entries.

The process of choosing a detector and updating C_R and C_R' is repeated until all valid detector templates have zero entries in D_R. Thus, at the end, for any template that is not in S, there are no more strings that have not been matched by a detector yet. In other words, the generated detectors cover all the unmatched strings that can possibly be covered as the algorithm keeps track of the number of nonself strings that have not been matched by any detector.

However, tuning a detector generation algorithm requires determining what is considered an optimal performance. If the goal is to minimize the number of detectors required for a fixed reliability level and the self-set, then the string length (l), m (the alphabet size), r (the matching threshold), and the matching rule for generating detectors need to be carefully chosen. In brief, the complexity of the detector generation process depends on the matching rule.

Singh (2002) extended the greedy algorithm to bigger alphabets. This is relevant in cases where the semantics of the information may get lost in binary representation. A lower number of false-positives were reported compared to results using binary representation.

4.3.4 Other Variants in Detector Generation

4.3.4.1 NSMutation Algorithm

NSMutation (Ayara et al., 2002) is a modified version of exhaustive algorithm, which introduces somatic hypermutation mechanism to improve performance. Particularly, instead of elimination of the candidate detector that matches the self-data, guided mutation is performed to attempt to make them valid detectors. The probability of mutation is considered as directly proportional to affinity between the candidate and self-sample. The difference in complexity comes from the time to mutate the matching region of length r of detector candidate. Mutation is limited to the region of length r, the upper bound of mutating is m_r. Matching threshold (r), detector lifetime rate, and mutation probability are major control parameters that can influence the performance of NSMutation algorithm. Generally, NSMutation is unique when

- Compared with exhaustive algorithm because NSMutation checks and solves the problem of redundant detector. Consequently, it reduces the total number of detectors.
- Compared with exhaust algorithm, it is tunable to balance between high coverage and efficient generation.
- Compared with linear, greedy, and binary template algorithm, which are highly restrictive to the *rcb* matching rule, NSMutation is more extensible.

4.3.4.2 Binary Template

In another approach, Wierzchon (2000) used binary string and an *rcb* matching rule to develop a deterministic algorithm to generate detectors that are more efficient in terms of minimal number of detectors (or receptors). Here, instead of generating a candidate detector randomly, it used a concept called "template."

A template is a string of length *l* over the alphabet {0, 1, *}, where *l* is the original length of the string, * stands for "do not care." Each template $t_{i,w}$ has the substring of length *k* starting on position *i* that equals a binary string *w* of length *k*, and the remaining bits are all *s. Note that *k* has the same meaning as *r* in *rcb* matching rule. The set of all possible templates, *T*, thus contains $(l - k + 1) \cdot 2^k$ different elements. *T* can be split into two disjoint subsets: T_S consisting of all the templates contained in at least one self-string and T_N, the set of remaining templates that are used to construct detector (receptor) strings. Typically, T_S is a low fraction of *T*. *T* can be represented as a matrix that has 2^k rows, one for each different *w*, and $(l - k + 1)$ columns, one for each starting point *i*. This notation makes it easier to analyze it numerically. Given a detector *r*, the number of unique strings from *U* detected by *r* is

$$D(l,k) = 2^{l-k} + (l - k) \cdot 2^{l-k-1} = 2^{l-k-1} \cdot (2 + l - k), \text{ for } k \leq \frac{l}{2}$$

This result can be extended from a binary alphabet to an alphabet of *m* symbols,

$$D_m(l,k) = m^{l-k} + (l - k) \cdot (m - 1) \cdot m^{l-k-1} = m^{l-k-1} \cdot [(l - k) \cdot (m - 1) + m], \text{ for } k \leq \frac{l}{2}$$

The discriminative power of a detector set, however, is not the sum of all $D(l, k)$s because different detectors can recognize common strings. Using a statistical approach, the average number of strings detected by *n* detectors is

$$d(l, k, n) = (1 - P_f(l, k, n)) \cdot 2^l$$

where $P_f(l, k, n)$ is the failure probability.

$$P_f(l, k, n) = (1 - P(l, k))^n \approx e^{-n \cdot p(l,k)}$$

where the approximation is valid for large n and small $P(l, k)$, the probability that two random strings match. For an "ideal" detector set, the coverage is fixed. For a given number of detectors that are not "ideal," detectors are chosen so that a maximum number of different templates are used. For implementation of the algorithm, a binary tree is used to represent the connection between the self-templates. Following the tree representation, all possible "self-strings" are reconstructed and those that are not self-strings can be found. These nonself, or undetectable, strings can come from two sources: strings that can be built from templates in T_S, or self-templates and those that have nonself templates (templates in T_N).

4.3.4.3   DynamiCS

A variation of NSA introduces "dynamic clonal selection algorithm (DynamiCS)" to deal with a nonself detection problem in a continuously changing environment (Kim and Bentley, 2002). In particular, DynamiCS is based on Hofmeyr's idea (Hofmeyr and Forrest, 2000) of dynamics of three different populations: immature, mature, and "memory detector" populations. Initial "immature detectors" are generated with random "genotypes." Using an NS, new immature detectors are added to keep the total number after a predefined number of generations ("tolerization period" T). If a detector is within its predefined "life span" L and the match counts are larger than a predefined "activation threshold" A, it becomes a memory detector. "Mature detectors" are used on all given antigens. However, a human security officer's confirmation (costimulation) is necessary to make the detector a memory detector, which makes the approach dependent on human interaction.

An enhanced NS algorithm (Hofmeyr, 1999) with multiple secondary representations was introduced to reduce the number of trials needed to generate detectors on the structured self as much as three orders of magnitude less. The suggested secondary representations included pure permutation, imperfect hashing, and substring hashing.

4.3.4.4   Schemata-Based Detection Rules

Hang and Dai (2004) introduced a new idea in detector generation by converting the data space into schemata space. Such a conversion compresses the data space. The problem space is n-dimensional vector space including categorical and numeric features. For real-valued features, a schema r is defined as the conjunction of the intervals as in the rules. Common schemata are those that are common in a group of rules. A number of common schemata are first evolved through a coevolutionary genetic algorithm (GA) in self-data space. The population used in the coevolutionary GA consists of a number of non-inter-breeding subpopulations. Species are initialized randomly, and new species are added into the population until the total number of species reaches a certain value. Then all the species are decoded into common schemata. Detectors are then constructed in the complementary space of the schemata using the traditional "generate-and-test" strategy. The candidate detector (detection rule) is rejected if it contains any common schemata.

4.4 Analysis of Negative Selection Algorithms

NSAs are based on the following argument (Forrest et al., 1994): given a reasonably large number of possible strings, the probability of two randomly chosen strings matching each other is relatively low; if the detector set is generated randomly and the abnormal strings can be considered as random to some extent, the probability that an abnormal string matches some detector increases with the number of detectors. Defining "failure probability" P_f as the probability that the detector set fails to detect a change, the main conclusion of Forrest et al. (1994) is

$$P_f \approx e^{-P_M N_R}$$

where P_M is the probability of a match between two random strings and N_R is the number of detectors in the detector set. P_M is largely decided by three major control parameters: m, the number of alphabet symbols; l, the length of string; and r, the number of contiguous bits *(rcb)* used in the matching rule.

4.4.1 Complexity of Detector Generation

D'haeseleer et al. (1996) analyzed an "exhaustive detector generating algorithm," and noted that a run finishes when the required number of detectors are generated; the number of detectors is chosen separately as a control parameter. The time complexity is

$$O\left(\frac{-\ln P_f}{P_m \cdot (1 - P_m)^{N_s}} \cdot N_S \right)$$

where P_M is the "matching probability," the probability that a randomly chosen string and a detector match, N_S is the size of the self-set. The space complexity is $O(l \cdot N_S)$. For specific matching rules, such as the dynamic programming approach (or linear time algorithm) with the *rcb* matching rule (D'haeseleer et al., 1996), its time complexity is $O((l - r) \cdot N_S) + O((l - r) \cdot 2^r) + O(l \cdot N_R)$ and space complexity is $O((l - r)^2 \cdot 2^r)$, where l is the string length; r is the number of contiguous bits in the matching rule *(rcb)*. This is more costly in space than the exhaustive algorithm. Also depending on the *rcb* matching rule, the greedy algorithm (D'haeseleer et al., 1996) has higher time complexity $O((l - r) \cdot 2^r \cdot N_R)$ and the same space complexity as the preceding algorithm, but it provides the maximum coverage for a given number of detectors, and the number of unmatched nonself strings is known (Balthrop et al., 2002).

Table 4.1   Complexity of Different Detector Generation Algorithm

NS Algorithm	Time Complexity	Space Complexity
Exhaustive approach (Forrest et al., 1994)	$O(m^l \cdot N_S)$	$O(l \cdot N_S)$
Dynamic programming (D'haeseleer et al., 1996)	$O((l - r + 1 \cdot N_S M^r) + O((l - r + 1) \cdot m^r) + O(l \cdot N_R)$	$O((l - r + 1)^2 \cdot m^r))$
Greedy algorithm (D'haeseleer et al., 1996)	$O((l - r + 1) \cdot N_S M^r) + O((l - r + 1) \cdot m^r \cdot N_R)$	$O((l - r + 1)^2 \cdot m^r))$
Binary template (Wierzchon, 2000)	$O(m^r \cdot N_S) + O((l - r + 1) \cdot m^r \cdot N_R)$	$O((l - r + 1) \cdot m^r) + O(N_R)$
NSMutation (Ayara et al., 2002)	$O(m^l \cdot N_S) + O(N_R \cdot m^r) + O(N_R)$	$O(l(N_S + N_R))$

Complexity of different negative detector generating algorithms (exhaustive, linear, greedy, binary template, and NSMutation) is summarized in Table 4.1 (taken from Ayara et al., 2002). The symbols used are as follows: N_S, number of self-data; N_R, number of mature detectors; N_{R0}, number of candidates; r, matching threshold; m, alphabet size ($m = 2$ for binary representation); and l, string length.

As shown in Table 4.1, the time complexities of the exhaustive algorithm and NSMutation are exponential with respect to the size of self. All the others have linear time complexity. However, when the matching threshold r approaches length l, the linear complexities may behave similarly to that of the exhaustive and NSMutation algorithms due to the exponential value m_r in their time complexities. NSMutation has a higher space complexity than that of the exhaustive algorithm. The linear, greedy, and binary template algorithms all have higher space complexity, although the binary template has the lowest among them. The greedy algorithm with higher alphabet, proposed by Singh (2002), has a time complexity of $O(m^r \cdot (l - r) \cdot N_R)$, where N_R is the prespecified number of detectors, and m the size of alphabet. The space complexity is $O(m^r \cdot (l - r)^2)$.

4.4.2   Immunological Holes

An important concept in detector generation is the notion of "immunological hole." An immunological "hole" corresponds to a set of nonself strings for which no antibody exists that fails to match the self. The concept of holes is applied to a large class of potential matching rules. For example, in the r-contiguous-symbols matching rule, if the self contains two strings ABC and $A'BC'$, where A, A', B, C, and C' are substrings, and B contains $(r - 1)$ symbols, then there is no antibody that will detect the nonself strings ABC' and $A'BC$. Thus, counting the number of holes under different conditions, determining how big the holes are in practice for a specific problem, and devising methods for detecting holes (e.g., by storing antibodies with different r values) are interesting problems.

The existence of holes imposes a lower bound on the failure probability P_f you can achieve with a detection method because it will always fail to detect holes (see Table 4.2). D'haeseleer et al. (1996) reported that the required number of detectors to achieve a certain acceptable P_f without taking holes into account the real P_f achieved with this detector set may be substantially higher than expected. Further, the failure probability associated with the holes themselves does not improve by distributing the algorithm if we use the same matching rule at all the sites.

D'haeseleer et al. (1996) gave an analysis of the number of detectors for a given failure probability, P_f, or the fraction of nonself strings that are not covered by the detector set. Denoting the information content (or entropy) of a self-set S of size N_S as $H(S)$, and the information about S that is missing in the detector set R as $H(S|R)$, it is concluded that the difference between $H(S)$ and $H(S|R)$, called "mutual information" of S and R, is

$$I(S; R) \equiv H(S) - H(S|R) \approx N_S \cdot log_2(1/P_f)$$

For string length l and alphabet size m, the lower bound of detector size

$$N_R \geq \frac{N_s \cdot log_2(1/P_f)}{l \cdot log_2(m)}$$

Table 4.2 Shows the Number of Holes and Best Achievable P_f for Different Configurations

L_S^a	N_S^b	L^b	r^b	P_M^c	Number of Holes[d]	Lowest Possible P_f^e
500 B	250	16	10	0.00391	634	0.0097
			9	0.00879	4,438	0.0677
			8	0.01953	21,076	0.3216
1 KB	250	32	11	0.00562	2,649	6.1676e-07
			10	0.01172	24,911	5.8000e-06
			9	0.02441	2,150,714	0.0005
			8	0.05078	5.1815e + 08	0.1206
	500	16	11	0.00171	882	0.0135
			10	0.00391	3,854	0.0588
			9	0.00879	24,937	0.3805

Note: These results were calculated on randomly generated self of sizes 500 bytes (B), 1 kilobytes (KB).
[a] Size of the dataset.
[b] Parameters chosen for the matching rule (rcb).
[c] Corresponding matching probability P_M.
[d] Number of holes present.
[e] Resulting best achievable failure rate P_f.
Source: Reported by D'haeseleer P., S. Forrest and P. Helman, An immunological approach to change detection: Algorithms, analysis and implications. *Proceedings of the 9th IEEE Computer Security Foundations Workshop*, pp. 18–26, Los Alamitos, CA, June, 1996.

Another lower bound for N_R can be given in terms of matching probability P_M, the probability that a string and a detector that are randomly chosen match each other according to the specific matching rule is

$$N_R \geq (1 - P_f)/P_M$$

It is shown that the holes are unavoidable. For example, in a partial matching rule *(rcb)*, two self-strings together may eliminate all detectors that are necessary to detect certain nonself strings. Given a string h and a matching rule M, if M has constant matching probability P_M, a self-set of size $N_S = |M'(h)| = P_M \cdot N_U$, where $M'(h)$ is the set of the detectors matching string h, always suffices to induce holes.

Esponda et al. (2003) have shown that the holes can be constructed by "crossover closure" (CC) method under various matching criteria. Accordingly, for a given set S of strings, and a fixed $1 \leq r \leq l$, applying the construction method on S, only holes can be constructed that are "crossed" combinations of bits in S. For example, assuming the string length, $l = 4$, $rcb = 2$, and for three self-strings $s_1 = 0110$, $s_2 = 1010$, and $s_3 = 1100$, it is possible to construct holes $h_1 = 1110$ and $h_2 = 0100$ as follows (Figure 4.4):

These examples have shown that the CC is a proper subset of the possible strings, and holes can be generated by the genetic algorithm's crossover operator for a set of bit strings, S. Also r-chunk matching rule can be viewed as a generalization of the *rcb* matching rule, which is related with the concept of CC.

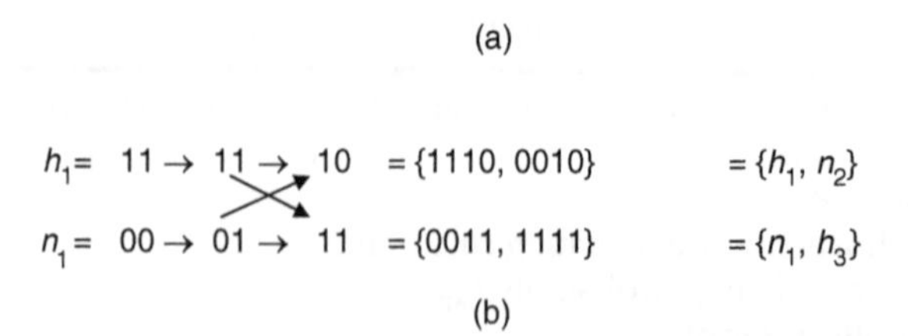

(a)

(b)

Figure 4.4 The figure illustrates the construction of holes from a given set of self using CC. (a) Holes ($h_1 = 1110$, $h_2 = 0100$) are constructed by $s_1 = 0110$, $s_2 = 1010$, and $s_3 = 1100$. (b) An additional hole h_3 can be constructed by the already found hole $h_1 = 1110$ and $n_1 = 0011$.

In another work, Stibor et al. (2004) developed a deterministic algorithm to generate all possible detectors using the r-chunk matching rule. Such a detector set is called a "perfect detector set," D_{perfect}, which contains a minimal number of detectors that recognize all elements in $U\backslash S$, where U is the representation space and S is the self-set (assuming all self-strings are included in S). The perfect detector set does not solve the problem of the holes. Instead, it clarifies the fact that holes are impossible to correct even when the complete self-set and proper matching rule are given. The algorithm uses a hash table H data structure to insert, delete, and search efficiently Boolean values, which are indexed with a composite key of r-chunk string concatenated with detector position p. It is divided into three phases. The total space size is $O(|\Sigma|^r)$. The time complexity of the entire three phases is

$$O((l-r)\cdot|\Sigma|^r) + O(|S|\cdot(l-r+1)) + O(|\Sigma|^r) = O(|\Sigma|^r)$$

The average number of generable detectors, depending on a self-set S, r-chunk length r, and alphabet size Σ was estimated as

$$\left(1 - \frac{1}{(l-r+1)\cdot|S|^r}\right)^{|S|\cdot(l-r+1)\cdot(l-r+1)} \cdot (l-r+1)\cdot|\Sigma|^r$$

However, it is hardly possible to have perfect coverage as for any string representation, but a matching rule with variable threshold may possibly fill the holes better.

4.5 Empirical Analysis: Binary Matching Rules and Detector Coverage

The original NS algorithm (described earlier) is very general and should work with any representation space and matching rule. It is clear that the algorithmic efficiency of generating good detectors varies with the type of representation space (continuous, discrete, hybrid, etc.), the detector representation, and the process that determines the matching (rule) ability of a detector.

Gonzalez et al. (2003a) analyzed and compared the effect of different binary matching rules (described in Chapter 3) in NS: r-contiguous matching, r-chunk matching, Hamming distance matching, and its variation "Rogers and Tanimoto (R&T) matching" to establish guidelines in selecting the matching rules for NSAs. Experiments were carried out on a two-dimensional space (unit square) using different encoding schemes: binary and Gray encoding, and observed the effect of matching rules on the area covered by individual detectors.

Experiments showed that the binary matching rules were not able to produce a good coverage of the nonself space. The r-chunk matching rule generated satisfactory coverage of the nonself space (Figure 4.5b); however, the self-space was covered by some lines resulting in erroneously detecting the self as nonself (false alarms). The Hamming-based matching rules generated an even more stringent

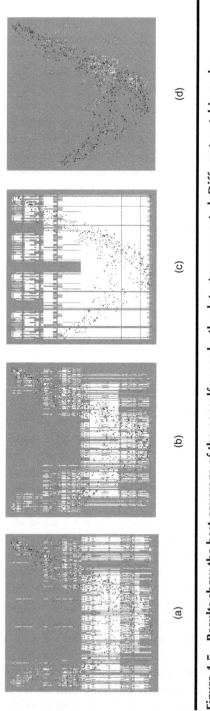

Figure 4.5 Results show the best coverage of the nonself space by the detectors generated. Different matching rules, parameter values, and encoding schemes (binary and Gray) were tested. The number of detectors is reported in (a) r-contiguous matching, $r = 9$, Gray coding; (b) r-chunk matching, $r = 8$, Gray coding; (c) r-chunk matching, $r = 7$, Gray coding; and (d) Hamming matching, $r = 13$, binary coding (same as R&T matching, $r = 10/16$).

result (Figure 4.5d) that covers almost the entire self-space. The parameter r, which works as a threshold, controls the detection sensitivity. A smaller value of r generates more general detectors (i.e., covering a larger area) and decreases the detection sensitivity. However, for a more complex self-set, changing the value of r from 8 (Figure 4.5b) to 7 (Figure 4.5c) generates a coverage with many holes in the nonself area, and still with some portions of the self covered by detectors. Therefore, this problem is not with the setting of the correct value for r, but a fundamental limitation of the binary representation that is not capable of capturing the semantics of the problem space. The performance of the Hamming-based matching rules is even worse; it produces a coverage that overlaps most of the self-space (Figure 4.5d).

In summary, advantages of string representation are (1) any data can be eventually represented in binary form; (2) it is easy to analyze; and (3) it is good for textual or categorical information. Its limitations include the comprehensibility problem (difficult to interpret in the original problem space), the potential scalability issue (string size and matching threshold value), and some difficulty in combining with other techniques (conventional algorithms, machine learning, etc.). Accordingly, the representation and matching rule for an NSA needs to be chosen in such a way that it accurately represents the data proximity in the problem space.

4.6　Real-Valued Negative Selection Algorithms

The real-valued representation encodes each data item as a vector of real numbers, where, the representation (self/nonself) space, U corresponds to a subset of R^n; only samples of one class are assumed. Specifically, such samples are considered to be representative data from the self-space. Then, based on these samples, a model of the self-set is built. For instance, the self-set can be considered to consist of all points within a certain distance from each sample point.

Different versions of real-valued negative selection (RNS) algorithms were proposed so far; they include

- A heuristic algorithm to generate "hyperspherical detectors"
- NS with detection rules (an evolutionary algorithm to generate the "hypercube detector"
- Randomized RNS (an algorithm for generating hyperspherical detectors using random process to optimize the distribution of detectors)
- V-detector algorithms
- NS with fuzzy detection rules (an evolutionary algorithm to generate "fuzzy rule detector")

Moreover, RNS algorithms can also be classified as (1) the "classical" generation-and-elimination strategy (Gonzalez, 2003); (2) evolutionary approaches, for example, GA (Dasgupta and Gonzalez, 2002); (3) one-shot randomized algorithm

(Ji and Dasgupta, 2004a and 2004b); or (4) optimization with aftermath adjustment (Dasgupta et al., 2004). These algorithms are described in the following text.

In RNS, a detector is defined by an n-dimensional vector that corresponds to the center and by a real value that represents its radius; therefore, a detector can be seen as a hypersphere in Rn. The detector-antigen matching rule is expressed by the "membership function" of the detector, which is a function of the detector-antigen Euclidean distance and the radius of the detector. This approach is similar to the NS greedy algorithm (D'haeseleer, 1995b), but in a real-valued representation space.

The input to the algorithm is a set of self-samples represented by n-dimensional points (vectors). The algorithm tries to evolve another set of points (called detectors) that cover the nonself space. This is accomplished by an iterative process that updates the position of the detector driven by the following two goals:

- Move the detector away from self-points
- Keep the detectors separated to maximize the covering of nonself space

The logical steps of the algorithm are shown in Figure 4.6, which are described in the pseudocode (NS Algorithm 4).

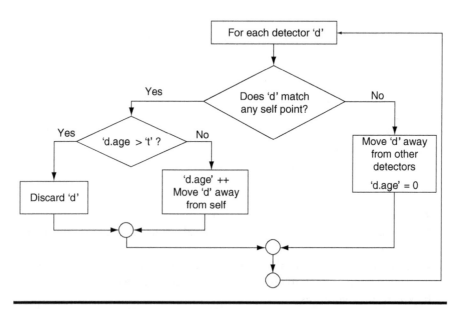

Figure 4.6 Shows an iteration of the RNS algorithm. This approach is similar to greedy algorithm but uses real-valued space. (From D'haeseleer P. An immunological approach to change detection: Theoretical results. *Proceedings of the 9th IEEE Computer Security Foundations Workshop*, pp. 18–26, Los Alamitos, CA, June 1996; D'haeseleer, P. *Further Efficient Algorithms for Generating Antibody Strings*, Technical Report CS95-3, The University of New Mexico, Albuquerque, NM, 1995a.)

The parameter r "specifies" the radius of detection of each detector of fixed size. Accordingly, for a new sample, s is detected by a detector d, if the distance between d and s is at most r. Because we do not want the detectors to match self-points, the shortest allowable distance for a good detector to the self-set is r. Therefore, the parameter r also specifies the allowed variability in the self-space.

To determine if a detector d matches a self-point, the algorithm calculates the k-nearest neighbors of d in the self-set. It then calculates the median distance of these k neighbors. If this distance is less than r, the detector d is considered to match self. This strategy makes the algorithm more robust to noise and outliers.

The function $\mu_d(x)$ is the matching function used for single the detector d. It indicates the degree of matching between x, an element of the self/nonself space, and d. It is defined as

$$\mu_d(x) = e^{-\frac{\|d-x\|^2}{2r^2}}$$

Each detector has an assigned age that is increased at each iteration, if it is inside the self-set. If the detector becomes old, that is, it reaches the maturity age t and has not been able to move out of the self-space, it will be replaced by a new randomly generated detector. The age is reset to zero when the detector is outside of the self-space.

The parameter η represents the size of the step used to move the detectors. To guarantee that the algorithm converges to a stable state, it is necessary to decrease this parameter at each iteration in such a way that $\lim_{i \to \infty} \eta_i = 0$. The following updating rule is used:

$$\eta_i = \eta_o e^{-i/\tau}$$

NS Algorithm 4 describes an RNS algorithm.

NS Algorithm 4: Real-valued-negative-selection, rns(r,η,t,k)
r radius of detection
η adaptation rate, i.e., the rate at which the detectors will adapt on
    each step
t once a detector reaches this age it will be considered to be mature
k number of neighbors to take into account

1 **while** stopping criteria is not satisfied
2   **for** each detector d do
3   *NearCells* ← k-nearest neighbors of d in the Self set
4   *NearCells* is ordered with respect to the distance to d
5   *NearestSelf* ←median of *NearCells*
6   If *dist (d, NearestSelf)* $< r$ Then
7           $dir \leftarrow \dfrac{\sum_c NearCells\,(d - c)}{\left|\sum_c NearCells(d - c)\right|}$

```
8           If age of d > t Then ▷ detector is old
9               Replace d by a new random detector
10    else
11          Increase age of d
12          d ← d + η. dir
13    endIf
14    else
15    age of d ← 0
```

$$
16 \quad dir = \frac{\sum_{d' \in Detectors} \mu_d(d')(d - d')}{\left|\sum_{d' \in Detectors} \mu_d(d')(d - d')\right|}
$$

```
17    d = d + η. dir
18          endIf
19    endFor
20 endWhile
```

4.6.1 Detector Generation Using Evolutionary Algorithms

Dasgupta and Gonzalez (2002) used a GA to evolve a set of rules (detectors) to cover the nonself space (Figure 4.7). The self-space consisted of a set S, a subset of $[0, 1]^n$; accordingly, a data point was represented as a feature vector $x = (x_1, ..., x_n)$ in $[0, 1]^n$.

A detection rule was considered to be good if it did not cover any positive sample (the self) and it covered a large area of the nonself space. A detector was then represented as a "detector rule" in the form

$$R_i : \textit{if} \, cond_i \, \textbf{then} \, \text{nonself, for } i = 1, ..., m$$

where $cond_i = X_1$ in $[low_1^i, high_1^i]$ and ... and X_n in $[low_n^i, high_n^i]$,

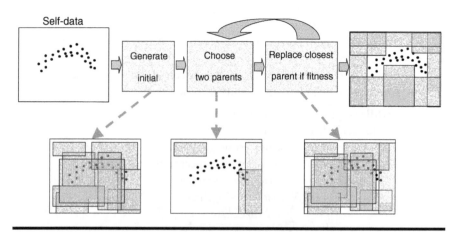

Figure 4.7   Generation of negative detection rules using a GA.

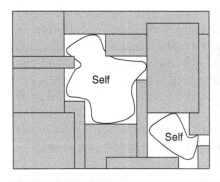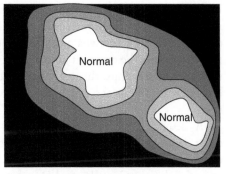

Figure 4.8 Self/nonself space. (a) Approximation of the nonself space by rectangular interval rules. (b) Levels of deviation from the normal in the nonself space.

where m is the number of detection rules and n the dimension of the Euclidean space, specifically, the unit n-dimensional. Thus, the condition part of each rule defines a hypercube on I^n. The evolution algorithm then evolved a set of rules and each chromosome encoded the condition part of a rule. Accordingly, in a two-dimensional space such detectors defined a rectangular shape as shown in Figure 4.8.

In this approach, a level of abnormality concept was introduced, which may be thought of as a degree of membership in the nonself set; instead of only two values to determine self or nonself, several discrete values were considered. Thus, a variability parameter v that represented the level of variability of a nonself point from normal (self) data was defined. Although v itself was not considered as part of the chromosome, different values of v were used to generate a hierarchical negative detection rules, which were grouped into different levels. Consequently, rules at level 1 contained rules at level 2, rules at level 2 contained rules at level 3, and so on. This is true because points within a certain distance s from self-samples can match rules at level k, whereas rule at level $k + 1$ can match self-points that are within a distance higher than s. Also, v can be interpreted as the radius of a hypersphere around each self-sample (Figure 4.8).

The fitness for a rule R was defined as

$$fitness(R) = volume(R) - \alpha \times num\text{-}self\text{-}samples(R)$$

where $num\text{-}self\text{-}samples(R)$ denotes the number of self-samples that match rule (detector) R, in other words, the number of self-samples that matches the subspace defined by rule R; $volume(R)$ denotes the volume of the hyperrectangle represented by R, mathematically,

$$volume(R) = \prod_{i=1}^{n} (high_i - low_i)$$

In this case, a new sample x was considered to match (satisfy) a rule R, denoted by $x \in R$ if the hypersphere with center at x and radius v intercepts the hyperrectangle defined by R. The parameter α denotes a coefficient of sensitivity, which for a specific rule determines the trade-off between the volume covered by it and its interception with the self-set. Thus, the total fitness is calculated as the sum of the fitness of all evolved rules minus the overlapping between the hyperrectangles defined by the rules.

A "sequential niching" algorithm was used to evolve a set of suitable rules to cover the nonself space (Dasgupta and Gonzalez, 2002). In this case, several runs of the evolutionary algorithm were needed, and a suitable rule was obtained at the end of each run. Gonzalez and Dasgupta (2002) used a "deterministic crowding (DC) niching technique," which performed better in covering the complementary space; thus, a smaller set of rules was obtained to estimate the amount of deviation more precisely. This new niching technique also allowed the generation of multiple rules in a single run instead of multiple runs as in "sequential niching." In the DC niching approach, a distance measure between two detectors c and p was defined as

$$dist(c, p) = \frac{volume(p) - volume(p \cap c)}{volume(p)}$$

where c and p are child and parent individuals, respectively. Note that this distance measure is not symmetric because it gives more importance to the area of the parent that is not covered.

The pseudocode for negative selection with detection rules (NSDR) algorithm using the DC are given in the following.

NS Algorithm 5: NS-detector-rules (S', *num_levels*,$\{v_1, ..., v_{numLevels}\}$)

Input: S: set of self samples

num_levels: number of deviation levels

$\{v_1, ..., v_{numLevels}\}$: allowed variability for each level

Output: Change Detection rules.

1 **For** i $= 1$ **to** *num_levels*
2   initialize population with random individuals
3 **For** j $= 1$ **to** num_gen
4   **For** k $= 1$ to pop_size/2
5   select two individuals, (*parent1, parent2*), with uniform probability and without replacement

6   apply crossover to generate an offspring (*child*)

7   mutate *child*

8   **If** *dist* (*child, parent*1) < *dist*(*child, parent*2) ∧ fitness(*child*) > fitness(*parent*1)

10   **Then** *parent*1←*child*

11   **ElseIf** *dist* (*child, parent*1)>= *dist*(*child, parent*2)

12   ∧ *fitness*(*child*) > *fitness*(*parent*2)

13   **Then** *parent*2 ← *child*

14   **EndIf**

15   **EndFor**

16 **EndFor**

17   extract the best individuals from the population and add them to the final solution

18 **EndFor**

4.6.2 Negative Selection with Fuzzy Detection Rules

Gonzalez (2003) applied fuzzy rules instead of crisp rules for detectors. That is, given a set of self-samples, the algorithm will generate fuzzy detection rules in the nonself space that can determine if a new sample is normal or abnormal. Results have shown that the use of fuzzy rules improves the accuracy of the method and produces a measure of deviation from the normal that does not need to partition the nonself space.

A fuzzy detection rule has the following structure:

$$\textbf{If } x_1 \in T_1 \wedge \ldots x_n \in T_n \textbf{ then } \text{non\_self}$$

where

    $(x_1, \ldots, x_n)$ = element of the self/nonself space being evaluated
    T_i = fuzzy set
    ∧ = fuzzy conjunction operator (in this case, *min*)

The fuzzy set T_i is defined by a combination of basic fuzzy sets (linguistic values).

Given a set of linguistic values $S = \{S_1, \ldots, S_m\}$ and subset $\hat{T}_i \subseteq S$ associated to each fuzzy set T_i,

$$T_i = \bigcup_{S_j \in \hat{T}_i} S_j$$

where $\bigcup$ corresponds to a fuzzy disjunction operator (here addition operator) defined as follows:

$$\mu_{A \cup B}(x) = \min\{\mu_A(x) + \mu_B(x), 1\}$$

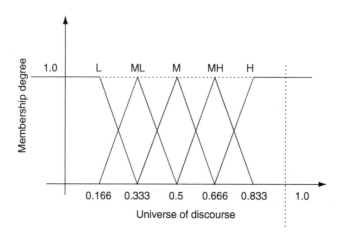

Figure 4.9 Partition of the interval [0, 1] in basic fuzzy sets.

An example of fuzzy detection rules in the self/nonself space with dimension $n = 3$ and linguistic values $S = \{L, M, H\}$ is as follows:

$$\textbf{If } x_1 \in L \wedge x_2 \in (L \cup M) \wedge x_3 \in (M \cup H) \textbf{ then non\_self}$$

Here, the basic fuzzy sets correspond to a fuzzy division of the real interval [0.0, 1.0] using triangular and trapezoidal fuzzy membership functions. Figure 4.9 shows an example of such a division using five basic fuzzy sets representing the linguistic values "low," "medium-low," "medium," "medium-high," and "high."

Given a set of rules $\{R^1, \ldots, R^k\}$, each one with a condition part $Cond_i$, the degree of abnormality of a sample x is defined by

$$\mu_{non\_self}(x) = \max_{i-1,\ldots,k} \{Cond_i(x)\}$$

where $Cond_i(x)$ represents the fuzzy true value produced by the evaluation of $Cond_i$ in x and $\mu_{non\_self}(x)$ represents the degree of membership of x to the nonself set; thus, a value close to 0 means that x is normal and a value close to 1 indicates that x is abnormal.

To generate the fuzzy rule detectors, the same evolutionary algorithm described in Section 6.1 (NSDR with DC) was used. However, the use of fuzzy rules does not require the generation of rules for different levels of deviation. Thus, all the detection rules are generated in a single run. However, the use of fuzzy rules requires changes to the chromosome representation, fitness evaluation, and distance calculation, which are discussed in the following text.

Each individual (chromosome) in the GA represents the condition part of a rule because the consequent part is same for all rules (i.e., the sample belongs to nonself). As described earlier, a condition is a conjunction of atomic conditions.

$S_1^1,..., S_m^1$	...	$S_1^n,..., S_m^n$
Gene 1		Gene n

Figure 4.10 Structure of the chromosome representing the condition part of a rule. Each gene represents an atomic condition $x_i \in T_i$ and each bit s_j^i is "on" if and only if the corresponding basic fuzzy set S_j is part of the composite fuzzy set T_j.

Each atomic condition, $x \in T_i$, corresponds to a gene in the chromosome that is represented by a sequence $(s_1^i, ..., s_m^i)$ of bits, where $m = |S|$ (the size of the set of linguistic values) and $s_j^i = 1$ if and only if $S_j \subseteq T_i$. That is, the bit s_j^i is "on" if and only if the corresponding basic fuzzy set S_j is part of the composite fuzzy set T_j. Figure 4.10 shows the structure of a chromosome which is $n \times m$ bits long (n is the dimension of the space and m the number of basic fuzzy set). Hamming distance was used as a distance measure. For example, if the s_j^i bit (see Figure 4.10) in both parent and child fuzzy rule detectors is set to 1, both individuals include the atomic sentence $x_i \in s_j$, that is, they use the jth fuzzy set to cover some part of the ith attribute. Then, the more bits the parent and the child have in common, the more common area they cover.

The fitness of a rule R^i is calculated by taking into account the following two factors: the fuzzy true value produced when the condition part of a rule, $Cond_i$, is evaluated for each element x from the self-set:

$$selfCovering(R) = \frac{\sum_{x \in Self'} Cond_i(x)}{|Self'|}$$

The fuzzy measure of the volume of the subspace represented by the rule:

$$volume(R) = \prod_{i=1}^{n} measure(T_i)$$

where "measure" (T_i) corresponds to the area under the membership function of the fuzzy set T_i.

The fitness is defined as follows:

$$fitness(R) = C \cdot (1 - selfCovering(R)) + (1 - C) \cdot volume(R)$$

where $C, 0 \leq C \leq 1$, is a coefficient that determines the amount of penalization that a rule suffers if it covers normal samples. The closer the coefficient to 1, the higher the penalization value (values between 0.8 and 0.9 were used).

The pseudocode in NS Algorithm 6 show the details of Negative Selection with Fuzzy Detection Rules (NSFDR) implementation; the time complexity of the algorithm is $O(num\_gen \cdot pop\_size \cdot |Self'|)$.

NS Algorithm 6: NS-fuzzy-detector-rules (*Self*)

Input: *Self*: set of self samples; Fuzzy Membership function

Output: A Fuzzy Rule set as Negative Detectors

```
1 initialize population with random individuals
2 for j = 1 to num_gen
3    for k = 1 to pop_size/2
4       select two individuals, (parent1, parent2), with uniform
         probability and without replacement
5       apply crossover to generate an offspring (child)
6       mutate child
7       if dist(child, parent1) < dist(child, parent2)
         ∧ fitness(child) > fitness(parent1)
8       then parent1 ← child
9       elseIf dist(child, parent1) ≥ dist(child, parent2)
10       ∧ fitness(child) > fitness(parent2)
11      then parent2 ← child
12      endIf
13   endFor
14 endFor
15 Take the better individuals from the population and add them to
   the detector set
```

4.6.3 Randomized Approaches in Generating (Fixed Size) Spherical Detectors

Gonzalez et al. (2003b) proposed a randomized approach based on "Monte Carlo integration" (Monte, 1995; Liu, 2001) to generate negative detectors. This approach assumed that all detectors had the same shape and size; particularly, hyperspheres of a fixed radius in an *n*-dimensional space were considered. Particularly, (1) Monte Carlo integration was used to estimate the volume of the self and nonself space, and it is also used to compute a rough estimate of the number of detectors needed to cover the nonself space; then (2) "simulated annealing" was used to optimize the detector distribution in the nonself space. Because the detectors are hyperspheres, the volume of the effective coverage of a detector was approximated as the volume of the inscribed hypercube:

$$V_d = \left(\frac{2r}{\sqrt{n}} \right)^n$$

Gonzalez et al. (2003b) assumed that the self-set $\hat{S}$ consisted of a collection of neighborhoods around each one of the self-sample points, S'; each neighborhood is defined as a hypersphere of radius r_s around a sample point. Therefore, the set $\hat{S}$ may be defined as

$$\hat{S} := \{x \in U : \text{there exists } s \text{ in } S', \|s - x\| \le r_{self}\}$$

and the volume of $\hat{S}$ is described as

$$V_{\hat{S}} := \int_U \chi_{\hat{S}}(x)dx$$

where

$$\chi_{\hat{S}} := \begin{cases} 1 & \text{if } x \in \hat{S} \\ 0 & \text{if } x \notin \hat{S} \end{cases}$$

Monte Carlo methods are well-established techniques with a strong mathematical foundation used for volume estimation, and they are being used here to estimate the coverage of a set of detectors. Also, this technique is useful in probabilistically estimating the overlap among detectors with different shapes, which otherwise will be cumbersome if a geometrical approach is followed.

Figure 4.11 illustrates the generation of hyperspherical detectors using Monte Carlo integration and simulated annealing.

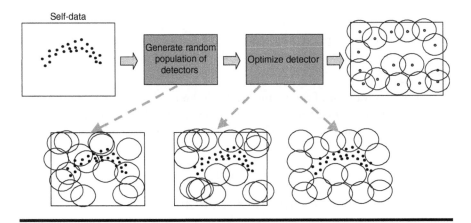

Figure 4.11 **A heuristic algorithm to generate hyperspherical negative detectors.**

4.6.3.1 Estimation of Detector Volume and Overlap

Let $X = [0, 1]^n$ be the system state space and $A \subseteq X$ a subset of X, whose volume needs to be computed. Also, a typical assumption is that it is hard to compute the volume of A analytically. If x is drawn from a uniform distribution on X, then $P(x \in A) = $ volume of A, denoted as $V(A)$ because $V([0, 1]^n) = 1$. Then the problem of computing $V(A)$ can be seen as the problem of estimating $P(x \in A)$.

Therefore, let U be a random variable uniformly distributed on X, denoted by $U \sim U(0, 1)$. Let $U_1, U_2, \ldots, U_n$ be a sequence of "independent and identically distributed (iid)" $U(0, 1)$ random variables. Then, the sequence $X_1, X_2, \ldots, X_N$ of random variables, generated as follows, are uniformly *iid* in $[0, 1]^n$, denoted by $X_i \sim U([0, 1]^n)$.

$$X_1 = (U_1, \ldots, U_n)$$
$$X_2 = (U_{n+1}, \ldots, U_{2n})$$
$$\ldots$$
$$X_N = (U_{(N-1)n+1}, \ldots, U_{nN})$$

To estimate the volume of A, generate a sequence $X_1, X_2, \ldots, X_N$, as defined earlier. Then, an estimation of the volume of A may be computed as

$$\hat{V}(A) = \frac{|\{i : X_i \in A\}|}{N}$$

where $|\cdot|$ denotes the number of points in a set. In other words, the volume of A is estimated as the fraction of points that lie in A. An estimate of the volume of A can also be expressed as

$$\hat{V}(A) = \frac{\sum_{i=1}^{N} Y_i}{N}$$

with $Y_i = I_A(X_i)$, where $I_A(\cdot)$ denotes the indicator function of set A. $Y_1, Y_2, \ldots, Y_N$ is a sequence of independent Bernoulli random variables, with $P(Y_i = 1) = P(X_i) = \int \ldots \int_A dx_1, \ldots, dx_n = V(A)$.

The main advantage of this method is that it is possible to calculate a confidence interval for the estimated volume $\hat{V}(A)$ as follows. To estimate the volume with a confidence of $(1 - \alpha)$, using the "Chernoff bound," it can be shown that if

$$N \geq \frac{3\ln(2/\alpha)}{\varepsilon^2 V(A)}, \text{ then}$$

$$P(|V(A) - \hat{V}(A)| > \varepsilon V(A)) \leq \alpha$$

However, as the dimensionality increases $V(A)$ approaches zero exponentially quickly, which will require a sample size exponentially large. Nevertheless, in many practical applications, dimensionality is such that reasonably small sample sizes are sufficient.

Monte Carlo integration was used to estimate $V_{\hat{s}}$ by generating m uniformly distributed random sample points and calculating an estimate of $V_{\hat{s}}$ as

$$V_{\hat{s}} \approx \hat{V}_{\hat{s}} = \frac{\sum_{i=1}^{m} \chi_{\hat{s}}(x_i)}{m}$$

where x_i, $i = 1, ..., m$, is the sequence of the sample points. The interval of confidence of such an estimate can be calculated for a given allowed error. Particularly, an interval of confidence equal to 0.998 was considered; therefore, the corresponding allowed error was calculated from the right-hand side of the inequality in the following equation:

$$P\left(|V_{\hat{s}} - \hat{V}_{\hat{s}}| < 3\sqrt{\frac{V_{\hat{s}} - \hat{V}_{\hat{s}}^2}{m}}\right) \approx 0.998$$

Once an estimate of the volume of self-space has been calculated, and an estimation of the effective coverage (volume) of a detector has also been computed, an estimation of the number of detectors necessary to cover the nonself space can be obtained.

Simulated annealing was used to find a good distribution of the detectors. The set of initial detectors is generated at random; subsequently, detectors are iteratively redistributed to approach the optimal distribution. This process was done to optimize an "objective function" $C(D)$ that measured the coverage of a set of detectors D. Thus, $C(D)$ is defined as

$$C(D) = Overlapping(D) + \beta \times SelfCovering(D),$$

where $D = \{d_1, ..., d_{num_{ab}}\}$ is a set of detectors (antibodies); num_{ab} is the number of detectors in D; $overlapping(D)$ is a function used to calculate the overlap between the detectors in D defined as

$$Overlapping(D) = \sum_{i \neq j} e^{\frac{-\|d_i - d_j\|^2}{r_{ab}^2}}, \quad i, j = 1...num_{ab}$$

and "SelfCovering" is a function that is used to "penalize" a detector when it matches any self-sample, and it is calculated as

$$SelfCovering(D) = \sum_{s \in S'} \sum_{d \in D} e^{\frac{-\|d - s\|^2}{((r_{ab} + r_{self})/2)^2}}$$

4.6.4 An Iterative Approach in Generating (Variably Sized) Negative Detectors

Dasgupta et al. (2004) used real-valued representation on $[0, 1]^n$, but it introduced a different way to generate hyperspherical negative detectors. This approach is similar to NSMutation (Ayara et al., 2002) but in real-valued space. At the beginning, an initial population of candidate detectors is generated at random. Such detectors then mature through an iterative process. In each iteration, the radius of each detector is calculated as $r_d = D - r_s$, where r_s is the variability around a self-point (see Figure 4.12a).

During an iterative process, detectors are moved away from self-input data and the other existing detectors. During this process, the detectors are ranked according to their coverage. The larger detectors are considered better fit and selected to go to the next generation. The smaller detectors are discarded and replaced with clones of the better-fit detectors. A clone of a detector is generated by moving the center of the original detector by a fixed distance to its proximity. In addition, new random detectors are introduced to explore new areas of the nonself space. The detector generation process terminates when a set of mature detectors that can provide significant coverage of nonself coverage are evolved.

Let $d = (c, r_d)$ be a detector. During the generation stage, the center of a detector is moved according to the following equation:

$$c^{new} = c + \alpha \frac{dir}{\|dir\|}$$

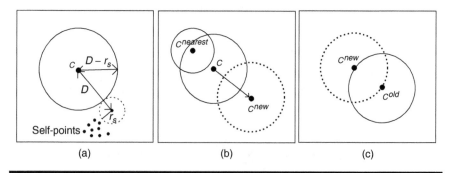

(a)   (b)   (c)

Figure 4.12 **The computational steps used during the detector maturation process are illustrated as follows: (a) shows a way to calculate and update the radius of a detector; (b) if a candidate detector overlaps with an existing detector (or self-points), then the candidate detector (i.e., its center c) is moved in the opposite direction to its nearest neighbor detector; (c) given a mature detector, a clone is created at a distance equal to its radius, and the direction where it is created is selected at random.**

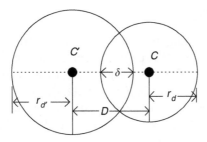

Figure 4.13 **The overlap between two detectors *d* and *d′* is computed in terms of the distance (*D*) between their centers (*c*, *c′*) and radii (r_d, $r_{d'}$).**

where $dir = c - c_{nearest}$ is the direction in which the center is moved; $c_{nearest}$ is the center of either the nearest detector or the nearest self-point (see Figure 4.12b). In contrast, cloning of a better-fit detector is described by

$$c^{clone} = c^{old} + r^{old} \frac{dir}{\|dir\|}$$

where c^{clone} is the center of a detector clone and c^{old} and r^{old} are the center and radius of the original detector, respectively. Because the detectors have hyperspherical shapes, overlap is necessary to cover the continuous nonself space (see Figure 4.12c). However, detectors are evaluated by the effective coverage, which is computed as their actual volume, but excluding overlap with other detectors (see Figure 4.13).

The overlapping measure of a detector *d* is computed as the sum of its overlap with the rest of detectors as

$$W(d) = \sum_{d \neq d'} w(d, d')$$

with

$$w(d, d') = (e^\delta - 1)^m \text{ and } \delta = \left(\frac{r_d + r_{d'} - D}{2r_d} \right)$$

$w(d, d')$ is the overlap measure between two detectors and *m* is the dimension of the feature space. This measure estimates the volume of the overlapped region.

Therefore, the following parameters are used during the detector generation process:

■ r_s = threshold value of a self-point (a point at a distance greater than or equal to r_s from a self-sample is considered to be as part of nonself)
■ α = parameter used to specify the offset when a detector is moved
■ ξ = maximum allowed overlap

4.6.4.1 Testing Process

The detection process is straightforward—the generated detectors are matched with new samples in test datasets. If a sample pattern, x is activated by (i.e., lies inside the recognition hypersphere of) a detector $y = (c, r)$, then an estimated distance from x to the self-set is computed as $\mu(x, y) = r\text{-}dist(x, c)$, where $dist(x, c)$ is the distance between sample pattern x and the center of detector y. Then, the degree of abnormality $A(x)$ of a matched pattern x is computed as the minimum of $\mu(x, y)$ among all activated detectors, y, that is,

$$A(x) = \min\{\mu(x, y)|\ y \text{ is activated by } x\}$$

Figure 4.14 shows the flow diagram of the iterative approach of generating variable-sized negative detectors.

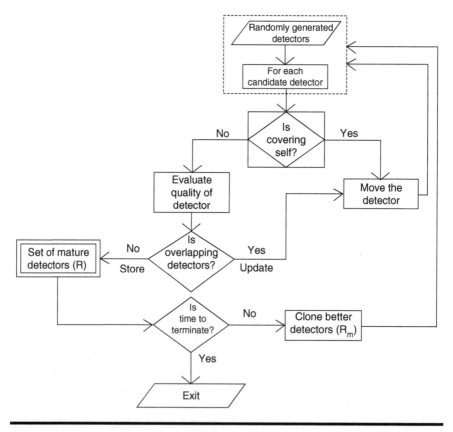

Figure 4.14 Flow diagram shows the steps of the variable-sized detector generation.

4.6.5   A Statistical Method: V-Detector Algorithm

Ji and Dasgupta (2004a, 2004b) proposed a new RNS algorithm called "V-detector," which has two main features as follows:

■ Variable-sized detector to achieve large coverage of nonself space with limited number of detectors
■ Estimation of the coverage using the generation process itself

This algorithm uses the conventional generate-and-test strategy of NS in a real-valued representation of data space on $[0, 1]^n$. Detector generation is performed by randomly generating a set of uniformly distributed random samples on I^n as possible centers of the detectors. If a point lies inside the self-set or it has already been covered by existing detectors, it is simply discarded. Also, the algorithm keeps track of the failed attempts to generate new detectors by such random points, which is further used to estimate the current coverage. If a new point either lies inside nonself or has not been covered by any detector, then it will give origin to a new detector given as a hypersphere whose radius is the maximum radius that will not make it match any self-sample. If the number of consecutive failed attempts that fall on a covered point reaches a limit m, the generation stage terminates with enough confidence that the coverage is sufficient. However, the value of m is not prespecified; it is rather decided by the estimated coverage as

$$m = \frac{1}{1 - \alpha}$$

where α is the current estimated coverage. The preceding equation may be explained as follows. If there is one uncovered point in a sample of size m', an estimate of the proportion of uncovered volume is $1/m'$, and the estimate of the coverage is then given by

$$\alpha' = 1 - \frac{1}{m'}$$

Actually, if there is no uncovered point in a sample of size m', there is a better than average chance that the actual coverage is larger than α'. Because m is decided by the earlier mentioned equation, after obtaining m consecutive points that are all covered, we can estimate that the actual coverage is very likely to be at least α.

An RNS with a constant size detector was used to compare it to the V-detector NS approach. The V-detector algorithm works well as long as the number of detectors is not taken into account. Overlap is not an important issue in the detector generation process as long as the new detector contributes to the coverage.

The main disadvantages of the V-detector approach are as follows:

■ The number of detectors is large compared to similar methods
■ Larger detectors do not necessarily contribute more to the coverage because they may overlap other detectors

NS Algorithm 7: Generate a negative detector set of V-detectors

Input S: set of self samples;

r_s: self radius

p: target coverage

ψ: significant level for hypothesis testing

T_{max}: maximum number of detectors

Output: A set of V-Detectors

1 $n \leftarrow \max\{5/p, 5/(1-p)\}$ {sample size required for hypothesis testing}
2 $D \leftarrow \varnothing$
3 **repeat**
4 $t \leftarrow 0$ {counter of "already covered" candidates}
5 $N \leftarrow 0$ {counter of valid candidates}
6 $C \leftarrow \varnothing$ {the collection of valid candidates}
7 $x \leftarrow$ random sample from $[0, 1]^n$
8 $r \leftarrow \infty$
9 **for all** s_i in S **do** {censored by self samples}
10 $d \leftarrow$ Euclidean distance between s_i and x
11 **if** $d \leftarrow r_s$ **then**
12   go to 7
13 **else**
14   $r \leftarrow \min\{r, d\}$
15 **end if**
16 **end for**
17 $N \leftarrow N + 1$
18 **for all** d_i in $D = \{d_i,\ i = 1, 2, \ldots.\}$ **do** {censored by existing detectors}
19   $d_d \leftarrow$ Euclidean distance between d_i and x
20   **if** $d_d < r(d_i)$ **then** {$r(d_i)$ is the radius of detector d_i}
21    $t \leftarrow t + 1$

22   $$z \rightarrow \frac{t}{\sqrt{np(1-p)}} - \sqrt{\frac{np}{1-p}}$$

23   **if z $>$ z then**
24   return D

```
25                          else
26                                    go to 31
27                          end if
28                  end if
29          end for
30 C ← C ∪ < x, r > {save a new candidate}
31 if N = n then
32          D ← D ∪ C
33          go to 4
34   else
35          go to 7
36 end if
37 until |D| = Tmax {Exception case: too many detectors to handle}
```

4.6.6 Multishaped Negative Detector Generation

This algorithm extends the model of RNS by incorporating multiple hypershape (hypersphere, hyperrectangle, or hyperellipse) detector representation in the unit hypercube $[0, 1]^n$ (Balachandran et al., 2007). These detectors are evolved applying a "structured genetic algorithm" (st. GA) with a niching technique for guiding the search. A st. GA is a particular form of evolutionary algorithm, which incorporates redundant genetic material controlled by a gene activation mechanism (Dasgupta and McGregor, 1994). It utilizes multilayered genomic structures (hierarchical chromosome) in which genes can be either active or passive (see Figure 4.15). An activation mechanism enables and disables the encoded genes, and high-level genes activate or deactivate sets of low-level genes. The redundancy is used to maintain genetic diversity to explore different areas of the parameter space.

In this work, a structured GA gene with a two-level representation is used, where the level 1 gene set holds the control information that either activates or

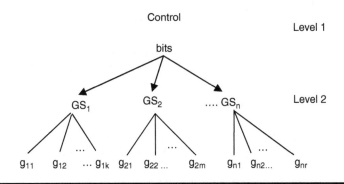

Figure 4.15 Generalized representation of a chromosome with *n* different gene sets.

Chromosome tree representation

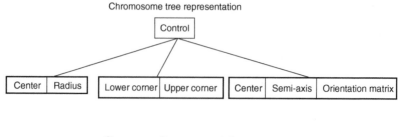

Chromosome linear representation

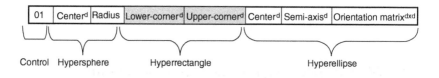

Figure 4.16 A single chromosome having a high-level control and low-level parameters encoding three different hypershapes: hyperspheres, hyperrectangles, and hyperellipses.

deactivates genes at level 2. Any gene set at level 2 can be expressed for fitness measure (shown in Figure 4.16).

Although each individual (chromosome) encodes multiple shapes, namely, hypersphere, hyperrectangle, and hyperellipse, it expresses only one shape in the "phenotypic space." Accordingly, "hypersphere genes" indicate the hypersphere *n*-dimensional center *c* and radius *R*, "hyperrectangle genes" hold information of the two points that specify the minimum and maximum coordinates in each dimension (e.g., the lower-left and upper-right corners of a rectangle in two dimensions), and "hyperellipse genes" contain its *n*-dimensional center ω, *n* semi-axes lengths l_i, and a square "orthonormal matrix V" of size $n \times n$, which specifies the orientation of the hyperellipse.

Thus, each detector shape, *d* in *n*-dimensional unit hyperspace is represented as

- Hypersphere, $d(c, r)$, with the center and the radius
- Hyperrectangle, $d[min_i, max_i]$ for each $i = 1, 2, \ldots, n$ dimension
- Hyperellipse, $d(p - \omega)^T A(p - \omega) < 1$, where $A = V\Lambda V^T$ and $\Lambda = (\lambda_{i,j})$ is an $n \times n$ diagonal matrix such that $\lambda_{ii} = 1/l_i^2$

Accordingly, a sample (with center *x* and variability *v*) forming the hypersphere $s(x, v)$ is considered matched (using some distance measure $dist(\cdot)$) by a detector, *d* of shape

- Hyper-sphere-shaped detector if $dist(c, x) < (v + r)$
- Hyper-rectangle-shaped detector if the circumscribed hypercube around the sample hypersphere intersects the detector

■ Hyper-ellipse-shaped detector if the closest point of the sample hypersphere to the center of the hyperellipse is inside it as shown in Figure 4.17 (Shapiro et al., 2005).

A number of runs of the evolutionary algorithm are required for a given self-set to generate a population of feasible detectors to cover the nonself space. Then, the best detector is selected to be added to the detector set, D. Figure 4.18 presents a pseudocode for the "evolutionary detector generation."

Each time a new detector is added to the detector list, the list is sorted in descending order based on coverage. The coverage is computed using a Monte Carlo estimation as follows: each detector is evaluated against a sequence of random points uniformly distributed in $[0, 1]^n$ to measure the percentage covered; the matched points are subsequently removed from the list. If any of the current detectors has an effective coverage of zero, it is assumed that the contribution of this detector is negligible and it is thus eliminated from the detector list.

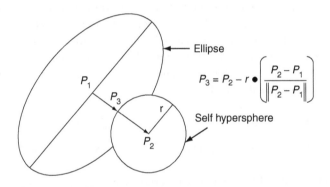

$$P_3 = P_2 - r \bullet \left(\frac{P_2 - P_1}{\|P_2 - P_1\|} \right)$$

Figure 4.17 **Determining the overlap of a hyperellipse detector with the self-set.**

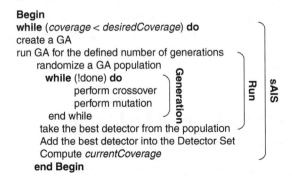

Figure 4.18 **An evolutionary algorithm to generate multishaped detectors.**

A detector, D provides a better coverage if the volume of D is as large as possible, while minimizing the overlap of D with S (the self-set) and the Detectors Set. Thus, the "fitness function" is defined as

$$fitness(D) = effective\text{-}coverage(D) - C(m) \times m$$

$$effective\text{-}coverage(D) = \hat{V}(D) - \beta$$

$$C(m) = \tan h\left(\frac{m}{\ln|S|}\right) \text{if } m > 0, \text{ and } C(m) = 0 \text{ if } m = 0$$

where $\hat{V}(D)$ is an estimate of the volume of D, β an estimation of the net overlap of D with the detector set, m the number of self-points that D matches, and $C(m)$ a "penalization factor" computed taking into account the number of self-points that D matches. The estimates of $\hat{V}(D)$ and β are obtained using a "Monte Carlo technique."

Balachandran et al. (2007) used tournament selection, two point crossover, and bit flip mutation in st. GA implementation. These operators are applied to each component of each gene of the chromosome. A specific mutation operator was used on the hyperellipse orientation matrix V: column vectors S^i and S^j, $1 \le i, j \le n$, are chosen at random, a real value θ is chosen from a "Gaussian distribution" with $\mu = 0$ and $\sigma = \pi/2$, and two components of vectors S^i and S^j to be mutated are picked at random and calculated as

$$S_i'^i = S_i^i \cos\theta + S_i^j \sin\theta \qquad S_i'^j = -S_i^i \sin\theta + S_i^j \cos\theta$$

$$S_j'^i = S_j^i \cos\theta + S_j^j \sin\theta \qquad S_i'^j = -S_j^i \sin\theta + S_j^j \cos\theta$$

Each run of the algorithm ends when certain coverage is reached, or when the algorithm fails to increase the coverage above a certain threshold for a certain number of iterations.

This work (Balachandran et al., 2007) showed a way to develop a unified framework for generating multishaped detectors in RNS algorithm. Results showed that multishaped detectors can provide better coverage compared to any single-shaped detectors. The uniform representation scheme and the evolutionary mechanism used in this work also serve as a baseline to include other shapes for efficient coverage of nonself space.

4.7 Applicability Issues of Real-Valued Negative Selection Algorithms

Real-valued representation seems appropriate if the underlying problem is continuous and can capture some continuous properties in the problem space. The paper by Ji and Dasgupta (2006) tries to clarify some issues raised on the applicability and weakness of NSAs, especially those in real-valued representation. Other than the general difficulties in learning algorithms, such as high dimensionality, there exist some issues with RNS, which are as follows:

■ Because the matching process in an NSA (or any learning algorithms) is built on the concept of affinity or distance, the results based on some converted discrete data may be fallacious. For example, the converted points will be distributed on separated (parallel) planes in the real space. The distance within one plane should not be interpreted in the same way as the distance between the planes. The connotation of being closer or farther apart is not the same as in the original data space. Therefore, the converted real-valued data not only fail to contribute to measure the distance or affinity between two points, they also limit the reasonable choice of a threshold for other fields in data.

■ The matching rule usually takes the form of a distance measure; selection of a specific matching rule should be according to the representation and detector shape.

■ Detector coverage depends on the interpretation of training data, which in most cases are incomplete (one-class classification problem). The statistical estimate of coverage using random sampling does not take the probability distribution of the data to be tested into consideration. Thus, the notion of enough coverage is always bias, which depend on how different the actual distribution is from uniform distribution. Detection rate, in contrast, depends on the actual distribution of test data.

Although the issues of NSAs' applicability is still an open debate, many difficulties reported in recent years are not related to the RNS algorithm itself. For example, the difficulty of high dimensionality, decision on optimal control parameters, and a good data model of the application domain are all important implementation issues for all methods.

4.8 Positive Selection (Detection)

In contrast to NS, "positive detection techniques" are widely used in pattern recognition, clustering, and other domains, where they generate a set of detectors that match self-points (instead of nonself points). In this case, a model of the self-set (training data) is used to classify a sample as part of either self or nonself. A simple model of a positive detection could be built using a nearest neighbor approach. If a point lies in a neighborhood of a sample self-point, then it will be labeled as belonging to the self-set (Figure 4.19).

Generally, a positive detector defines the neighborhood by assuming a hypersphere with a certain radius centered on each of the self-points. Moreover, detectors can be defined in a more sophisticated way by using some clustering algorithm

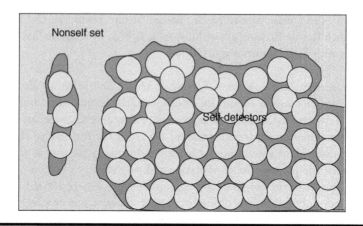

Figure 4.19   PS approaches. The goal of PS is to cover the self-set with an appropriate set of detectors.

on the self-sample points. Therefore, a sample point can be classified as belonging to a cluster by measuring its distance to it. A measure of the distance from a sample to a cluster may be defined in terms of the Euclidean distance to the "cluster centroid." Another way to define such distance is a "normalized Euclidean distance" defined as

$$dist() = \frac{\|s - K\|}{\sigma_K}$$

where K is the cluster centroid and σ_K the standard deviation that represents the sparseness of the cluster.

A basic positive characterization approach can be defined using a nearest neighbor criterion. A "crisp characterization," will classify a feature vector as normal or abnormal. However, a "noncrisp characterization" can be considered by introducing degrees of "abnormality," represented by a values in the interval [0, 1], with 1 indicating "normal" and 0 "abnormal." Thus, a function $\mu_{nonself}$ defined on $[0, 1]^n$ associate to each feature vector, measures the distance to the nearest self-sample point. In this case, no additional model of the self-space was considered; the self-sample points were considered as the definition of the self-sub-space. Thus, $\mu_{nonself}$ is defined as

$$\mu_{nonself}(x) = D(x, self) = \min\{d(x, s) : s \in self\}$$

where $d(x, s)$ denotes the distance between x and a self-sample point s. Thus, $D(x, self)$ is the distance from x to the closest point in the self-sample set. Other distance measures such as a "Minkowski metric" can also be used.

Dasgupta and Gonzalez (2002) compared the negative characterization approach to a positive characterization method ("*kd*-tree"). Although the positive characterization approach gave more precise results, it was more costly in time and space when compared to the negative characterization technique. In another work, Gonzalez (2003) made a comparison of "hybrid neuro-immune system" (HNIS) and "self-organizing maps" (SOM) for anomaly detection application.

Ji and Dasgupta (2006) used "support vector machine" (SVM) as the positive detection technique and compared with the V-detector algorithm. SVM appears to have provided more generalization, but the performance primarily depends on a suitable choice of the kernel function.

Both PS and NS can be a reasonable choice based on various reasons. An application with a large amount of self-data seems appropriate for NS.

4.9 Negative Database

The negative database (NDB; Esponda et al., 2004; Esponda, 2005) is one of latest developments in NSAs. Accordingly, the "self" is a database that stores a collection of data records referred to as positive database (PDB), and the complementary database, NDB, constitutes all possible records that are not in DB. The interesting property of this representation concerns the difficulty of inferring DB from a given NDB, thus, enhancing the privacy of the sensitive information without any encryption.

4.9.1 Negative Database Representation

Given an arbitrary set of strings l-length defined over {0, 1, *}, where * is the "don't care" symbol, determining which strings are not represented in NDB is an NP-hard problem. As all the possible strings that are not in DB constitute the NDB, NDB = $U -$ DB, where U denotes the universe defined over the same alphabet (Esponda et al., 2004). A string s is in DB if and only if it fails to match all entries in the NDB (Table 4.3).

Table 4.3   Example of a DB, Its Corresponding U-DB, and a Possible NDB Representing U-DB

DB (Self)	(U-DB)	NDB (Nonself)
010	000	*00
101	001	0*1
	011	*11
	100	
	110	
	111	

4.9.2 Representation of Negative Databases as Satisfiability Problems

Esponda et al. (2004, 2007a) show that there is a natural mapping from NDBs to the satisfiability (SAT) problems. For example, an instance of 3-SAT can be represented as the instance of NDB, and this can be done in the following manner:

1. The Boolean formula can be written in conjunctive normal form (CNF) and can be defined over some variables or literals (e.g., x_i).
2. This Boolean formula can be mapped to NDB, where each clause corresponds to a negative record in NDB. Each variable or literal in the clause can be represented as
   a. 1 if the literal or variable is appeared as negated
   b. 0 if the literal or variable is appeared as unnegated (x_i)
   c. * if the literal is not appeared in the clause
3. The resulting expression returns true for every assignment that is not in the PDB, and false for each truth assignment.

In the preceding example (Table 4.4), there are five literals or variables: x_1, x_2, x_3, x_4, x_5. If any literal is present in the clause or Boolean formula, then 0 or 1 is assigned in the NDB depending on the negation of variable or unnegated variable. In particular, if the variable is unnegated then there will be 0 in the NDB; in contrast, if the variable is negated, then there is 1 for that variable in NDB. But if the variable is not present in the formula, then an asterisk "*" is assigned for that variable. Therefore, the total number of bits in each of the NDB record depends on the number of literals. In the first formula (Table 4.4), x_3 and x_4 are not present; therefore, in the NDB, "*" is assigned in their positions. Also, x_1 and x_2 are not negated, therefore 0 is assigned, whereas x_5 is in the negated form, therefore 1 assigned for it.

4.9.3 Approaches to Generate Negative Databases

Esponda et al. (2004) proposed two algorithms for the creation of NDBs: (1) a deterministic algorithm and (2) a randomized algorithm. In particular, the first

Table 4.4 Representation of NDB as CNF

Boolean Formula	NDB
(x_1 or x_2 or $\bar{x}_5$) and	00**1
($\bar{x}_2$ or x_3 or x_5) and	*10*0
(x_2 or $\bar{x}_4$ or $\bar{x}_5$) and	*0*11
($\bar{x}_1$ or $\bar{x}_3$ or x_4)	1*10*

approach is a monolithic model for NDB construction, which is called the prefix algorithm; and the other is called the randomized algorithm, which is a more complex approach for the NDB creation.

4.9.3.1   Prefix Algorithm

The prefix algorithm is an iterative way of building longer prefixes of assigned values, and demonstrates a simple construction of NDBs having the property of isomorphism. Isomorphism is an ideal goal for NDBs because it guarantees no false-positives, whereas most probabilistic generators have some tolerance to false-positives. False-negatives (incorrectly classifying data that is in the positive dataset) should be avoided by definition. This algorithm produces NDB of size $O(l\,|DB|)$ where l is the length of each record in bits (Figure 4.20).

4.9.3.2   Randomized Algorithm

Although the prefix algorithm given in Figure 4.20 is very simple and generates compact NDB, the weakness of this algorithm is that any single positive record will have its counterpart represented in the last record of the NDB with only the last bit complemented. In larger PDBs, there are no proven hardness properties, and the simplicity of the algorithm makes it easy to reverse in many cases.

The randomized algorithm by Esponda et al. (2004) presents more complex steps for NDB creation. It is a more realistic problem creator because it is probabilistic and does not follow any direct sequence of decisions for generating NDBs. The algorithm's complexity relative to the prefix algorithm leads to the intuition of being

Prefix algorithm
Let w_i denote an i-bit prefix and W_i a set
of i-length bit patterns.
1. $i \leftarrow 0$
2.   Set W_i to the empty set
3.   Set W_{i+1} to every pattern not present in
      DB's w_{i+1} but with prefix in W_i
4.   for each pattern V_p in W_{i+1}{
5.   Create a record using V_p as its prefix
          and the remaining positions set to the
          don't care symbol.
6.   Add record to $N\,D\,B$.}
7.   Increment i by one
8.   Set W_i to every pattern in DB's w_i
9.   Return to step 3 as long as $i < l$.

Figure 4.20   The prefix algorithm. (From Esponda F., S. Forrest and P. Helman. *Enhancing Privacy through Negative Representations of Data.* UNM Computer Science Technical Report TR-CS-2004-18, March 2004.)

"hard to reverse," but this specific algorithm does not provide any properties that can be used to prove the empirical performance of any given NDB (Figure 4.21).

In Figure 4.21, the main loop generates all prefixes that are not represented by the PDB similar to the prefix algorithm. It then uses a *Pattern_Generate* function to randomize the wildcards and assign positions in the bit positions at the end of the prefix. The algorithm can generate every possible entry that is not a positive record. An example of NDBs generated by the preceding two algorithms is shown as follows (Table 4.5):

Randomize_*NDB* algorithm
Let w_i denote an i-bit prefix and W_i a set
of i-length patterns.
1. $i \leftarrow [\log_2(l)]$
2.   Initialize W_i to the set of every pattern
     of i bits.
3.   Set W_{i+1} to every pattern not present in
     DB's w_{i+1} but with prefix in W_i
4.   for each pattern V_p in W_{i+1} {
5.   Randomly choose $1 \leq j \leq l$
6.   for $k = 1$ to j do {
7. $V_{pg} \leftarrow$ Pattern_Generate $(\pi(DB), V_p)$
8.   Insert V_{pg} in *NDB*.}}
9.   Increment i by one
10. Set W_i to every pattern in *DB*'s w_i
11. Return to step 3 as long as $i < l$.

Figure 4.21   Algorithm B, the randomized NDB algorithm. This generates an NDB that is intuitively difficult to reverse, but it is not tunable and provides no guarantee of "hardness." (From Esponda F., S. Forrest and P. Helman. *Enhancing Privacy through Negative Representations of Data.* **UNM Computer Science Technical Report TR-CS-2004-18, March 2004.)**

Table 4.5   Column 1 Gives an Example of DB and Column 2 Gives NDB Generated by Prefix Algorithm and Column 3 Gives NDB Generated by Randomized Algorithm

DB	Prefix NDB	Randomized NDB
0011	10**	10**
0101	000*	*00*
1100	011*	011*
1111	0010	0*10
	0100	0*00
	1101	1*01
	1110	**10
		*110
		*010

Esponda et al. (2006) proposed a distributed NDB model, which not only fixed some properties of the monolithic database but also created some new ones of its own. The main disadvantage of the model is the size of the database. Because one database stores one record, the size of the composite database grows as a linear function of the number of positive records.

4.9.4 Operations on Negative Database

Basic database operations such as initialization, insertion, and deletion are used to describe the NDB. The proposed algorithms for these operations may cause the size of the NDB to grow unreasonably. It is important for any implementation to control the number of entries that match a particular string. Some operations can be described as follows (Esponda et al., 2006):

Insert operation. There are two inputs for this operation: NDB and the string x (to be inserted) and outputs an NDB′ database that matches every string matched by NDB and every string matched by x.

Delete operation. There are two inputs for this operation: NDB and the string x (to be inserted) and outputs an NDB′ database that matches every string matched by NDB and removes every string matched by x.

Morph operation. This operation takes NDB as input and outputs NDB′ of same binary string but both NDB and NDB′ are different as some records of NDB are not in NDB′ and vice versa. This allows NDBs to have in different representations for the same data.

4.9.4.1 Negative Algebra

There are many algebraic operations for NDB such as relational algebra operations for sets. These operations include negative select, negative union, cartesian product, join and intersection, negative cartesian product, and negative join. Some of these operations are described as follows (see Esponda et al., 2007b for details):

Negative intersection ($\bar{\cap}$). This is just the opposite of the intersection operation on the sets, that is, this operation uses De Morgan's law and will union the elements of both the NDB (Esponda et al., 2007b). The negative intersection can be defined as follows:

$$NDB_3 = NDB_1 \mathbin{\bar{\cap}} NDB_2 = \{x | x \in NDB_1\} \cup \{y | y \in NDB_2\}$$

Table 4.6 illustrates the example of negative intersection ($\bar{\cap}$).

Negative union ($\bar{\cup}$). This operation is opposite of the intersection operation for sets, where only those strings or records which are common to both

Table 4.6 Negative Intersection of NDB$_1$ and NDB$_2$

NDB$_1$	NDB$_2$	NDB$_3$ = NDB$_1$ $\overline{\cap}$ NDB$_2$
10**	0*10	10**
011*	1*01	011*
000*	*010	000*
		0*10
		1*01
		*010

Table 4.7 Negative Union of NDB$_1$ and NDB$_2$

NDB$_1$	NDB$_2$	NDB$_3$ = NDB$_1$ $\overline{\cup}$ NDB$_2$
10**	0*10	0110
011*	1*01	1001
000*	*010	

Table 4.8 Negative Join ($\bowtie$) of NDB$_1$ and NDB$_2$

NDB$_1$	NDB$_2$	NDB$_3$ = NDB$_1$$\bowtie$NDB$_2$
10**	0*100	10***
011*	1*011	011**
000*	*0100	000**
		0*100
		1*011
		*0100

input NDBs are placed in the resultant NDB, and can be defined as follows:

$$\text{NDB}_3 = \text{NDB}_1 \overline{\cup} \text{NDB}_2 = \{z | z = x \odot y, \quad xMy, \quad (x \in \text{NDB}_1 \wedge y \in \text{NDB}_2)\}$$

where

Coalesce $x \odot y$ = Two strings x and y of length n coalesce into string z iff x matches y and for all $1 \le i \le$

Match xMy: Two strings, x and y, match iff for all i $((x[i] = y[i]) \vee (x[i] = *) \vee (y[i] = *))$

Therefore, when the two strings in both databases are exactly the same (comparing each bit of the string, i.e., xMy), then this string is appended into the NDB$_3$, for example (Table 4.7).

Negative join. This operation results in the NDB, which contains all the strings except those that are in the join of two PDBs: DB$_1$, DB$_2$. An example of negative join is given in Table 4.8.

In summary (as described by Esponda et al., 2006), a properly designed NDB can act as a privacy-preserving storage system, which has the following properties:

- *Hard to reverse.* Given an NDB, there should be no algorithm for obtaining the positive-image PDB that is more efficient than exhaustive search.
- *Singleton NDB.* Each hard-to-reverse entry in NDB represents either a string in PDB, or no string at all, that is, reversing the database does not introduce "false"-positive entries.
- *Easy to update.* There should be efficient algorithms for adding and deleting entries from the PDB.
- *Obfuscated size.* The size of the positive-image PDB should not be visible from the NDB.
- *Probabilistic.* A particular binary string s that belongs to PDB should have many possible representations in NDB.

However, Danezis et al. (2007) described an efficient implement cryptographic hash function to achieve the same functionalities as NDBs with security guaranteed.

4.10 Summary

This chapter discusses various elements of NSAs in detail (Ceong et al., 2003; Dasgupta, 1999a,b; Esponda and Forrest, 2002; Kim and Bentley, 2001; Stibor et al., 2005, 2006). Different NSAs are characterized by their representation schemes, matching rules, and detector generation processes. The detector generation mechanism in NSAs, as described in the original model (Forrest et al., 1994), is a randomized algorithm that generates candidates and then eliminates those that match self-samples or training data. Except for the difference in the matching rules developed later, most NSAs using string representation have the same or similar detector generation process. In contrast, a few deterministic generation algorithms were also designed. In many cases, they were described so as to study the algorithmic complexity and detector coverage analytically (Ayara et al., 2002; D'haeseleer et al., 1996; Wierzchon, 2000). Because string representations provide a more convenient platform for such analysis, deterministic algorithms are often discussed in such representations. Kaers et al. (2003) categorized major detector generation algorithms into two types: those built heavily on the assumption of the string representation: linear, greedy, and binary template and those relatively independent of the "antibody morphology": exhaustive and NSMutation.

The NSAs' uniqueness and strength can be grouped into two levels. The fundamental level includes some features that make this method really special:

- No prior knowledge of nonself is required (D'haeseleer et al., 1996).
- It is inherently distributable; no communication between detectors is needed (D'haeseleer et al., 1996).

- It can hide the self-concept (Esponda et al., 2004). At the other level, it has various strengths that may not be totally unique to this method.
- Compared with other change detection methods, NSAs do not depend on the knowledge of defined "normal." Consequently, checking activity of each site can be based on a unique signature of each while the same algorithm is used over multiple sites.
- The quality of the check can be traded off against the cost of performing a check (Forrest et al., 1994).
- If the process of generating detectors is costly, it can be distributed to multiple sites because of its inherent parallel characteristics.
- Detection is tunable to balance between coverage (matching probability) and the number of detectors (D'haeseleer, 1996).

Some limitations of the string representation in NSA are as follows:

- Binary matching rules are not able to capture the semantics of some complex self/nonself spaces.
- It is not easy to extract meaningful domain knowledge.
- In some cases, a large number of detectors are needed to guarantee a desired level of detection (Scalability issue).
- It is difficult to integrate the NS algorithm with other immune algorithms.
- Crisp boundary of self and nonself may be hard to define.

In real-valued representation, detectors are represented by hypershapes in an n-dimensional space. The algorithms use geometrical heuristics to distribute the detectors in a uniform way on the nonself space.

Some limitations of the real-valued representation in NSA are as follows:

- The issue of geometrical shapes
- The handling of dimensionality
- The estimation of coverage
- The selection of distance measure

Ji and Dasgupta (2006), in a recent study, pointed out that NSAs are not appropriate to be used as a general classification method because they use samples from one class in training. There exist inconsistency in the terminology and the assumptions used in NSAs to clarify the confusion or misunderstanding in applicability (Freitas and Timmis, 2003). Compared with binary or string representation, formal analysis is needed for the genre of real-valued representation.

NSAs play an important role in the research of artificial immune systems (AIS). Chronological development of NSA and its variants are reported by Ji and Dasgupta (2007).

4.11   Research Questions

1. Study the characteristics for matching rules.
2. How hard is it to find detectors that do not match self with different representations and matching rule?
3. How many detectors are required providing reasonable protection?
4. How does the number of detectors scale with size of self?
5. How to pick l, r, N_R, and other parameters in binary NS algorithm?
6. Develop a suitable test suite for evaluating NS algorithms?
7. How to address various applicability issues of real-valued NS algorithms?
8. Develop a real-world application of NDB. Compare the results with other methods.
9. Design and implement a new NDB algorithm.
10. Use various negative operations (algebra) to implement an efficient NDB.

4.12   Review Questions

1. What are different NSAs? Which is the main goal of these algorithms?
2. What are the main features of the NS process that occurs in biological immune systems?
3. Which immune natural process is simulated by artificial NS?
4. Write a pseudocode showing the major steps of an NSA.
5. List some real-world applications where NSAs could be used.
6. Why only samples from one class are sufficient for an NSA?
7. What are the main stages in an NSA?
8. Classify the following statements into either describing detector generation phase or monitoring/testing/detection phase:
   - A set of detectors are generated by some randomized process.
   - Candidate detectors that match any of the self-samples are eliminated, whereas a subset of the remaining ones is kept.
   - Matching rules are based on T cell/antigen affinity measures.
   - Detectors are used to check whether new incoming patterns correspond to self or nonself instances.
   - If an input pattern matches a detector, then it is identified as part of nonself.
9. Give a specific problem and representation where a binary representation is a better choice than a real-valued vector representation.
10. What are the advantages of binary representation over real-valued vector representation, if any?
11. What are the advantages of real-valued vector representation over binary representation, if any?

12. What is the major limitation of the random generation approach in the original description of the NSA?
13. What are the two phases of the NS dynamic programming approach described in the chapter?
14. Given $l = 4$, $r = 2$ and $S_1 = 0\ 0\ 0\ 0$, $S_2 = 1\ 0\ 0\ 1$, $S_3 = 0\ 1\ 1\ 0$, and $S_4 = 0\ 1\ 0\ 1$? Could you list the nonself strings as possible detectors?
15. What are the time and space complexity of the dynamic programming approach in negative detector generation?
16. What are the advantages of the greedy algorithm over the dynamic programming approach?
17. What does the following statement mean?
    "obtain a better coverage of the string space"
18. Analyze the following statements:
    ■ Generating detectors at random will take exponential time in the set of the size of the sample self-set (S) and r matching
    ■ The dynamic programming approach will run in linear time in the size of S and the number of detectors, but grows exponentially in r
    ■ The greedy approach will also run in linear time in the size of S and the number of detectors, and grows exponentially in r
19. Explain the notion of immunological hole. Why is it an important concept in detector generation?
    ■ Illustrate with examples (self-set) where holes exist?
    ■ Illustrate how the holes can be generated using crossover closure for a given set of self-strings.
20. Explain the main idea behind RNS algorithms.
21. Explain the main ideas behind the following approaches:
    ■ Detector generation using an evolutionary algorithm
    ■ Randomized NS
    ■ Fuzzy NS
    ■ V-detector NS
    ■ NSA
    ■ Multishaped NS
    ■ Combining NS and classification techniques
22. Mention some applicability issues of binary NSA. Mention some applicability issues of real-valued NSA.
23. What is an NDB? Mention different NDB algorithms and compare them.
24. Illustrate with an example to show the equivalence between the NDB and CNF. How does it relate to SAT problem?
25. Mention different operations used in NDB. Illustrate with an example negative algebra, for example, negative union and negative join.
26. What are the desired properties of an NDB design to use as a privacy-preserving storage system?

References

Ayara, M., J. Timmis, R. de Lemos, L. de Castro and R. Duncan. Negative selection: How to generate detectors. In Timmis J. and Bentley P. J. (Ed), *Proceedings of the 1st International Conference on Artificial Immune Systems (ICARIS)*, vol. 1, pp. 89–98, 2002.

Balachandran, S., D. Dasgupta, F. Nino and D. Garrett. A framework for evolving multi-shaped detectors in negative selection. *IEEE Symposium on Foundations of Computational Intelligence (FOCI)*, pp. 401–408, Honolulu, April 1–5, 2007.

Balthrop, J., F. Esponda, S. Forrrest and M. Glickman. Coverage and generalization in an artificial immune system. *Proceedings of the Genetic and Evolutionary Computation Conference (GECCO 2002)*, pp. 3–10, Morgan Kaufmann Publishers, New York, July 2002.

Ceong, H. T., Y. I. Kim, D. Lee and K. H. Lee. Complementary dual detectors for effective classification. *Proceedings of Second International Conference on Artificial Immune System (ICARIS 2003)*, pp. 235–253, Edinburgh, U.K., 2003.

Coutinho, A. The self non-self discrimination and the nature and acquisition of the antibody repertoire. *Ann. Immunol. (Inst. Past.)*, 131D, 1980.

Danezis, G., C. Diaz, S. Faust, E. K¨asper, C. Troncoso and B. Preneel. Efficient negative databases from cryptographic hash functions. In Garay J. A. (Ed), *Information Security Conference 2007*, Springer, Valparaiso, Chile, 2007.

Dasgupta, D. An anomaly detection algorithm inspired by the immune system. In Dasgupta D. (Ed), *Artificial Immune Systems and their Applications*, Springer, New York, pp. 262–277, 1999a.

Dasgupta, D. An overview of artificial immune systems and their applications. In Dasgupta D. (Ed), *Artificial Immune System and Their Applications*, Springer, New York, pp. 3–23, 1999b.

Dasgupta, D. and F. Gonzalez. An immunity-based technique to characterize intrusion in computer networks. *IEEE Trans. Evol. Comput.*, 6(3), 1081–1088, 2002.

Dasgupta, D., K. KrishnaKumar, D. Wong and M. Berry. Negative selection algorithm for aircraft fault detection. *Proceedings of Third International Conference on Artificial Immune Systems (ICARIS 2004)*, pp. 1–13, Catania, Italy, 2004.

Dasgupta, D. and D. R. McGregor. A more biologically motivated genetic algorithm: The model and some results. *Cybern. Syst.: Int. J.*, 25(3), 447–469, 1994.

D'haeseleer, P. *Further Efficient Algorithms for Generating Antibody Strings*. Technical Report CS95-3, The University of New Mexico, Albuquerque, NM, 1995a.

D'haeseleer, P. *A Change-Detection Algorithm Inspired by the Immune System: Theory, Algorithms and Techniques*. Technical Report CS95-6, The University of New Mexico, Albuquerque, NM, 1995b.

D'haeseleer, P. An immunological approach to change detection: Theoretical results. *Proceedings of the 9th IEEE Computer Security Foundations Workshop*, pp. 18–26, Kenmare, Ireland, June 1996.

D'haeseleer, P., S. Forrest and P. Helman. An immunological approach to change detection: Algorithms, analysis and implications. *Proceedings of the 1996 IEEE Symposium on Research in Security and Privacy*, IEEE Computer Society Press, Los Alamitos, CA, pp. 110–119, 1996.

Esponda, F. *Negative Representations of Information.* Dissertation, The University of New Mexico, 2005.

Esponda, F., E. S. Ackley, P. Helman, H. Jia and S. Forrest. Protecting data privacy through hard-to-reverse negative databases. In Katsikas S. K., Lopez J., Backes M., Gritzalis S., and Preneel B. (Eds), *ISC, Volume 4176 of Lecture Notes in Computer Science*, pp. 72–84, Springer, 2006.

Esponda, F., E. S. Ackley, P. Helman, H. Jia and S. Forrest. Protecting Data Privacy through Hard-to-Reverse Negative Databases. *Int. J. Inform. Security (IJIS)*, 6(6), 403–415, 2007a.

Esponda, F. and S. Forrest. *Detector Coverage Under The r-contiguous Bits Matching Rule*, The University of New Mexico, Albuquerque, NM, TR-CS-2002-03, 2002.

Esponda, F., S. Forrest and Helman P. The crossover closure and partial match detection. *Proceedings of the 2nd International Conference on Artificial Immune Systems (ICARIS)*, pp. 249–260, Edinburgh, U.K., 2003.

Esponda, F., S. Forrest and P. Helman. *Enhancing Privacy through Negative Representations of Data.* UNM Computer Science Technical Report TR-CS-2004-18, March 2004.

Esponda, F., E. D. Trias, E. S. Ackley, and S. Forrest. *A Relational Algebra for Negative Databases.* The University of New Mexico, Computer Science Technical Report TR-CS-2007-18, November 2007b.

Forrest, S., A. S. Perelson, L. Allen and R. Cherukuri. Self-nonself discrimination in a computer. *Proceedings of the 1994 IEEE Symposium on Research in Security and Privacy*, IEEE Computer Society Press, Los Alamitos, CA, 1994.

Freitas, A. A. and J. Timmis. Revisiting the foundation of artificial immune systems: A problem-oriented perspective. *Proceedings of Second International Conference on Artificial Immune System (ICARIS)*, LNCS 2787, pp. 229–241, Springer, Edinburgh, U.K., 2003.

González, F. *A Study of Artificial Immune Systems Applied to Anomaly Detection.* PhD. Dissertation, The University of Memphis, May 2003.

Gonzalez, F. and D. Dasgupta. An immunogenetic technique to detect anomalies in network traffic. *Proceedings of the Genetic and Evolutionary Computation Conference (GECCO)*, New York, July 9–13, 2002.

Gonzalez, F. A. and D. Dasgupta. Anomaly detection using real-valued negative selection. *Genet. Progr. Evol. Machines*, 4, 383–403, 2003.

Gonzalez, F., D. Dasgupta and J. Gomez. The effect of binary matching rules in negative selection. *Proceedings of the Genetic and Evolutionary Computation Conference (GECCO 2003)*, LNCS 2723, pp. 195–206, Chicago, IL, July 2003a.

Gonzalez, F., D. Dasgupta and L. F. Nino. A randomized real-value negative selection algorithm. *Proceedings of Second International Conference on Artificial Immune System (ICARIS 2003)*, Edinburgh, U.K., September 2003b.

Hang, X. and H. Dai. *Constructing Detectors in Schema Complementary Space for Anomaly Detection, Lecture Notes in Computer Science*, LNCS 3102, *Proceedings of Genetic and Evolutionary Computation Conference (GECCO 2004)*, pp. 275–286, Springer, Seattle, Washington, 2004.

Hofmeyr, S. A. *An Immunological Model of Distributed Detection and its Application to Computer Security.* PhD thesis, The University of New Mexico, 1999.

Hofmeyr, S. A. and S. Forrest. Architecture for an artificial immune system. *Evol. Comput.*, 7(1), 45–68, 2000.

Ji, Z. and D. Dasgupta. Augmented negative selection algorithm with variable-coverage detectors. *Proceedings of 2004 Congress on Evolutionary Computation (CEC 2004)*, pp. 1081–1088, Portland, OR, June 2004a.

Ji, Z. and D. Dasgupta. Real-valued negative selection algorithm with variable-sized detectors. *Proceedings of the Genetic and Evolutionary Computation Conference (GECCO)*, LNCS 3102, pp. 287–298, Portland, OR, 2004b.

Ji, Z. and D. Dasgupta. Applicability issues of the real valued negative selection algorithms. *Genetic and Evolutionary Computation Conference (GECCO)*, (Received Best Paper Award). Seattle, Washington, July 2006.

Ji, Z. and D. Dasgupta. Revisiting negative selection algorithms. *Evolutionary Computation Journal*, Issue 15.2, July 2007.

Kaers, J., R. Wheeler and H. Verrelst. The effect of antibody morphology on non-self detection. *Proceedings of Second International Conference on Artificial Immune System (ICARIS 2003)*, Edinburgh, U.K., 2003.

Kappler, J., N. Roehm and P. Marrack. T Cell tolerance by clonal elimination in the Thymus. *Cell*, (49), 273–280, 1987.

Kim, J. and P. J. Bentley. An evaluation of negative selection in an artificial immune system for network intrusion detection. *Proceedings of the Genetice and Evolutionary Computation Conference (GECCO 2001)*, San Francisco, CA, 2001.

Kim, J. and P. Bentley. Immune memory in the dynamic clonal selection algorithm. *Proceedings of the 1st International Conference on Artificial Immune Systems (ICARIS)*, pp. 59–67, Canterbury, U.K., September 2002.

Liu, J. S. *Monte Carlo Strategies in Scientific Computing*. Springer, 2001.

Monte, Computational Science Education project. Introduction to Monte Carlo Methods. Available at http://en.wikipedia.org/wiki/Monte_Carlo_method, 1995.

Shapiro, J. M., G. B. Lamont, and G. L. Peterson. An evolutionary algorithm to generate hyper-ellipsoid detectors for negative selection. *Proceedings of the Conference on Genetic and Evolutionary Computation*, vol. 1, ACM Press, Washington, pp. 337–344, 2005.

Singh, S. P. N. Anomaly detection using negative selection based on the r-contiguous matching rule. *1st International Conference on Artificial Immune Systems (ICARIS)*, University of Kent at Canterbury, September 9–11, 2002.

Stibor, T., K. M. Bayarou C. Eckert. An investigation of R-Chunk detector generation on higher alphabets. *Proceedings of the Conference on Genetic and Evolutionary Computation (GEECO)*, vol. 1, pp. 299–307, Seattle, Washington, 2004.

Stibor, T., P. Mohr, J. Timmis and C. Eckert. Is Negative Selection Appropriate for Anomaly Detection? *Proceedings of the Genetic and Evolutionary Computation Conference (GECCO)*, Washington, June 25–29, 2005.

Stibor, T., J. Timmis and E. Claudia. The Link between r-contiguous Detectors and k-CNF Satisfiability. *Proceedings of IEEE World Congress on Computational Intelligence (Special Session on Recent Development In Artificial Immune Systems) in Congress on Evolutionary Computation*, Vancouver, July 17–21, 2006.

Tax, D. M. J. *One-Class Classification*. PhD thesis, Technische Universiteit, Delft, 2001.

Wierzchon, S. Discriminative power of the receptors activated by k-contiguous bits rule. *J. Comput. Sci. Technol.*, 1(3), 1–13, 2000.

Chapter 5

B Cell–Inspired Algorithms

This chapter describes clonal selection algorithms and artificial immune networks (AINs), which are mainly inspired by B cells' response to antigens. First, the main features of clonal selection algorithms and their similarities to evolutionary approaches are presented. Then, continuous and discrete immune network (IN) models are discussed. Finally, different versions of IN model are described briefly.

5.1 Clonal Selection Algorithms

Clonal selection algorithms are developed based on the clonal selection theory (Burnet, 1959) proposed nearly 50 years ago. The main immunological elements used are

- Maintenance of a specific memory set
- Selection and cloning of most stimulated antibodies
- Removal of poorly stimulated or nonstimulated antibodies
- Affinity maturation (hypermutation) of activated immune cells
- Generation and maintenance of a diverse set of antibodies

Clonal selection algorithms (De Castro and Von Zuben, 2000), however, are very similar to a kind of evolutionary algorithm; namely, evolutionary strategies (Beyer and Schwefel, 2002), although they have a different biological inspiration. Clonal selection algorithms are also population-based search and optimization algorithms generating a memory pool of suitable antibodies for solving a particular problem.

In clonal selection algorithms, each antibody and antigen is represented by a set of attributes $\{x_1, x_2, \ldots, x_n\}$. Thus, antibodies and antigens may be represented as either n-dimensional points in a metric space such as Euclidean space or use binary encoding of the attributes; however, other representations are also used.

The antigenic affinity of each antibody is typically defined based on a metric, usually, the Euclidean distance. Also, some operators are defined to introduce genetic variation to the antibodies based on their antigenic affinities. First, a cloning operator is defined to make exact copies (clones) of those antibodies having higher antigenic affinities; the higher the antigenic affinity, the higher the number of clones an antibody can generate. Then some genetic variation is introduced to these antibodies (through a mutation operator) to allow them for better matching with the antigens.

Although several variations of clonal selection algorithms have been introduced, most algorithms have similar features as that of the basic clonal selection algorithm (De Castro and Von Zuben, 2000) presented in Figure 5.1 (the different steps of the flow diagram are shown in Figure 5.2).

During the affinity maturation process, when mutated antibodies are added to the current population to reselect the best individuals and keep them as the memory of current antigen, it is necessary to compute the affinities of the new antibodies toward the antigen; therefore, the whole set of antibodies need to be ranked, and subsequently, a selection process needs to be performed.

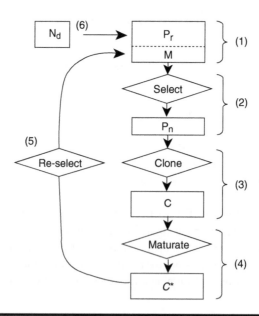

Figure 5.1   Generic clonal selection algorithm.

clonal selection algorithm( )

1. Initialization

Create initial random population of antibodies *P*

2. Antigenic presentation

while not stopping criterion met *do*

    for each antigen *do*

        2.1. Affinity evaluation

        present current antigen to current antibody

        population *P*

        for each antibody *do*

        compute antibody affinity with current antigen

> **2.2. clonal selection and expansion**
>
>   select a subset of antibodies in *P* with highest
>
>   affinities
>
>   generate exact copies (clones) of antibodies with
>
>   highest affinities according to their affinities,
>
>   the higher the affinity, the higher the number of
>
>   clones
>
> **2.3. affinity maturation**
>   mutate all clones at a rate inversely proportional to
>
>   their affinities with current antigen
>
>   add mutated antibodies to *P* and re-select the best
>
>   individuals to keep them as the memory of current
>
>   antigen
>
> **2.4. metadynamics**
>   replace antibodies with lowest affinities by randomly
>
>   generated new antibodies

Figure 5.2   Description of the generic clonal selection algorithm.

Table 5.1   Clonal Selection Algorithms versus Evolutionary Algorithms

Features	Evolutionary Algorithm	Clonal Selection Algorithm
Search space	Set of chromosomes	Set of antibodies
Candidate solution, individuals	Chromosome	Antibody
Individual representation	Any (strings, real vectors, etc.)	Any (strings, real vectors, etc.)
Population size	Fixed	Fixed
Fitness function (performance measure)	Fitness based on the function	Affinity
Operators	Chromosome selection Mutation Crossover	Clone selection Hypermutation

In Table 5.1, the generic clonal selection algorithm is compared with a basic evolutionary algorithm. It is to be noted that differences primarily lie in the terminology used. In clonal selection algorithms, the mechanism to select the fittest antibodies is based on their affinities with the antigens. Therefore, traditional selection mechanisms used in evolutionary computation, such as proportionate selection or tournament selection are easily adapted to be used in clonal selection algorithms. In evolutionary algorithms, typically, the probability that an individual be selected is determined by its fitness. The case is the same for the antibody or antigen affinities.

Several versions of clonal selection algorithms (ClonAlg) are proposed by De Castro and Von Zuben (2000) and De Castro (2003) and are used to perform pattern recognition and multimodal function optimization tasks. The algorithm is depicted in Figure 5.3.

There are different versions of clonal selection algorithm; the version of clonAlg applied to pattern recognition problems assumes a set of patterns to be recognized as input, whereas the version used for optimization assumes an objective function to be optimized.

5.2   Immune Network Models

The basic IN models (Anderson et al., 1973; Neuman, 1992; Vertosick and Kelly, 1989) tried to model network properties of immune cells in the absence of foreign antigens. These immune networks are mostly considered as idiotypic networks (Burnet, 1957; De Boer and Hogeweg, 1989). Generally, an antibody could be represented as a pair (p, e), where p is the antibody's collection of paratopes and e the set of epitopes. Each antibody has two paratopes and two epitopes, which are the specialized parts of the antibody that identify and are identified by other molecules, respectively.

clonAlg( )
1) Randomly choose an antigen Ag_j and present it to all *Abs* in the repertoire **AB**.
2) Determine the vector f_j that contains the affinity of Ag_j to all the *Abs* in **AB**.
3) Select the *n Abs* from **AB** with the highest affinity, which will form a new set $\mathbf{Ab}^j_{\{n\}}$ of high-affinity in relation to Ag_j.
4) The *n* selected *Abs* will be cloned (reproduced) independently and proportionally to their antigenic affinities, generating a repertoire $\mathbf{C}^j$ of clones: the higher the antigenic affinity, the higher the number of clones generated for each of the selected *Abs*.
5) The repertoire $\mathbf{C}^j$ is submitted to an affinity maturation process, which mutates antibodies inversely proportional to the antigenic affinity, generating a population of mature clones: the higher the affinity, the smaller the mutation rate.
6) Determine the affinity f_j^* of the mature clones $\mathbf{C}^{j*}$ in relation to antigen Ag_j.
7) From this set of mature clones $\mathbf{C}^{j*}$, reselect the one with highest affinity ($\mathbf{Ab_j}^*$) in relation to Ag_j to be a candidate to enter the set of memory antibodies ($\mathbf{Ab}_{\{m\}}$). If the antigenic affinity of this Ab in relation to Ag_j is larger than its respective memory Ab, then $\mathbf{Ab_j}^*$ will replace this memory Ab.
8) Finally, replace the lowest affinity Ab's from $\mathbf{Ab}^j_{\{r\}}$, in relation to Ag_j, by new individuals in.

Figure 5.3 The clonal selection algorithm (clonAlg). (From De Castro, L. N. and F. J. Von Zuben, in *Proceedings of Genetic and Evolutionary Computation Conference (GECCO) 2000, 36–37.*)

These models consider interactions between antibodies and antigens as complementary antigen–antibody matches or among antibodies themselves; however, exact matching between antibody and antigen (i.e., between paratope and epitope) is not used. Instead, some antibody–antigen matching measure (complementary measure) is defined such that if its value is below some threshold, the antibody does not react to the antigen at all. Binding between an antigen and an antibody depends on how well an antibody's paratopes match an antigen's epitope; the closer this match, the stronger the bind. Also, a similarity or affinity measure needs to be defined to measure how well two antibodies match.

In case of AINs, antibodies and antigens need to be explicitly defined, which is called a shape–space. Other elements are also to be considered in an AIN model, which include

1. *Number of antigen epitopes.* Some models consider one epitope, others consider several epitopes.
2. *Number of antibody epitopes.* Some models do not consider any epitope at all, others consider one or several epitopes.
3. *Number of antibody paratopes.* Some models consider one or two paratopes.
4. *Types of binding interaction between antibodies.* Some models consider only paratope–paratope interaction, only paratope–epitope or both paratope–paratope interactions and paratope–epitope binding.

When an antibody has one paratope, it can bind to one epitope at a time; but, when it has two paratopes, one paratope can bind to an antigen, whereas the other may bind to an antibody with a similar epitope to the antigen.

Most AIN applications start with an input dataset that corresponds to a set of antigens stimulating an immune network, which goes through a dynamic process, until it reaches stability. Depending on the application, either the concentration of each type of antibodies or the structure of the AIN or both are used as results.

All models assume an initial configuration of the IN; in some cases, the initial configuration of the IN is produced at random. The IN undergoes a stimulation process caused by the set of foreign antigens to the network; however, some models consider analyzing the IN's intrinsic behavior when no antigens are present.

An IN is also represented as a graph, where nodes, edges, and arrows represent antibodies, the interactions among them, (see Figure 5.4) and stimulation to the AIN by foreign antigens, respectively.

IN models (De Boer, 1989) can be classified into two categories: continuous- and discrete models. In continuous models, the immune response is assumed to be continuous, as opposed to discrete models where it is in discrete time steps. Continuous models are described by a set of differential equations, and its purpose is mainly in modeling biological phenomenon. However, discrete models are typically abstract functional models, which are called AINs, and its purpose is to solve real-world computational problems.

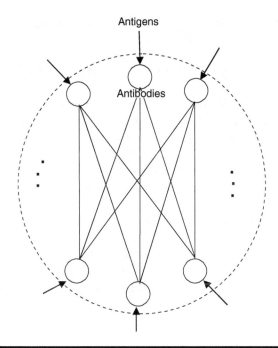

**Figure 5.4   Illustration of an IN. Nodes, edges, and arrows represent antibod-
ies, interaction among them, and stimulation to the AIN by foreign antigens,
respectively.**

5.2.1   Continuous Immune Network Models

Continuous IN models are defined as a set of differential equations that attempt
to predict the concentration of a finite number of antibodies in the IN at a certain
time, during or after an immune response. Models based on differential equations
do not focus on the structure of the IN, but on the antibody and foreign antigen
concentration, although they assume that all antibodies interact with one another
and the antigens interact with all antibodies as well.

The change in the concentration of a specific antibody is represented as the sum
of two terms:

$$dx_i/dt = \Delta x_i = \text{internal network dynamics} + \text{antigen driven dynamics} \quad (5.1)$$

where x_i is the concentration of type i antibody at a given time. The first term
models the interaction among antibodies, that is, the effect of stimulation and
suppression of antibody paratopes by other antibody epitopes; this term describes
natural death of antibodies. In contrast, the second term models the stimulation of
antibodies by antigens.

Also, in continuous models, a set of equations are used to predict the concentration of specific antigens and is defined as

$$dy_i/dt = \Delta y_i = \text{antigen elimination} + \text{natural antigen birth or death} \quad (5.2)$$

where y_i is the concentration of antigen of type i at a given time t.

Therefore, one IN model can be distinguished from another by noticing how each one of these terms represents the dynamic interaction processes. Different models are also characterized by the way they represent antibodies, including the number of paratopes, epitopes, and antigens.

A more specific form of Equation 5.2 may be given by

$$\Delta x_i = \text{internal interactions} - \text{antibody damping} + \text{antigen driving} \quad (5.3)$$

where the first term represents the natural dynamics of the idiotypic network as a result of antibody–antibody interactions; the second term models the reduction of cells in the absence of stimulation by antigens; whereas the third term represents the antigenic effects.

5.2.1.1 Jerne's Idiotypical Network

Jerne's (1974) model (Weisbuch et al., 1990) introduced the following equation to describe the change of lymphocytes of a certain type:

$$\frac{dx_i}{dt} = x_i \sum_{j=1}^{N} f(E_j, K_j, t) - x_i \sum_{j=1}^{N} g(I_j, K_j, t) + k_1 - k_2 x_i \quad (5.4)$$

where the first term represents the total stimulation of lymphocytes of type i by excitatory signals as a sum of excitatory signals received from stimulating lymphocytes. Accordingly, $f(E_j, K_j, t)$ is a measure of excitatory signals from idiotypes in E_j on a type i lymphocyte at time t. K_j is a constant associated with the strength of the affinity between lymphocyte of type i and idiotypes in E_j. In a similar fashion, the second term expresses the total effect of inhibitory signals from other lymphocytes on a lymphocyte of type I; thus, I_j expresses a lymphocyte whose combining sites recognize idiotypes on type I cells. In addition, k_1 is the rate at which type i lymphocytes enter the network and k_2 is a natural death rate of type i lymphocytes in the absence of antigen.

In this model, a differential equation describes the change in the concentration of lymphocytes of each type. Thus, the network presents a dynamic behavior even in the absence of stimulating antigens. To describe the dynamic behavior of a foreign antigen, an additional term needs to be included to represent the interaction of corresponding type i lymphocyte with external antigens.

5.2.1.2   Coutinho and Varela's Idiotypical Network

In Coutinho and Varela's model (Coutinho, 1993; Varela and Coutinho, 1991), both idiotypes bound to a cell surface and free antibodies are considered. The concept of network sensitivity for an idiotype is introduced and defined as a function of the affinity between such idiotype and the network antibodies. Here, m_{ij} denotes the affinity between two idiotypes of type i and j. The network sensitivity for the ith idiotype is denoted by $\sigma_i(t)$ and it is thereby defined as

$$\sigma_i(t) = \sum_{j=1}^{N} m_{ij} f_j(t) \tag{5.5}$$

$f_j(t)$ is the amount of free idiotypes of type i. After a maturation process, specific B cells generate free antibodies. Thus, a differential equation that describes the change in the concentration of free antibodies is defined as

$$\frac{df_i(t)}{dt} = k_i b_i(t) mat(\sigma_i(t)) + k_2 f_i(t)\sigma_i(t) - k_3 f_i(t) \tag{5.6}$$

$b_i(t)$ is the number of idiotypes attached to the surface of B cells.

$$\frac{db_i(t)}{dt} = k_4 b_i(t) prol(\sigma_i(t)) + meta[i] - k_5 b_i(t) \tag{5.7}$$

Mat() is the lymphocyte maturation function. *Prol* regulates the probability of proliferation, and *meta[i]* represents the metadynamics that results from adding resting lymphocytes into the active IN. Both *mat*() and *prol*() functions are considered as bell-shaped.

5.2.1.3   Farmer, Packard, and Perelson's Idiotypical Network

This IN model considers the microdynamics of the antibodies and antigens interaction. In this case, the model keeps track of the proportions of each type of antibodies among the population. An antibody is considered as a pair of paratope and epitope (p, e), which are explicitly represented by binary strings (Farmer et al., 1986).

The affinity measure takes into account all possible matching by shifting antibody j. Each shift of antibody i is matched against antibody j. Therefore, the matching affinity m_{ij} between antibodies i and j is defined as

$$m_{ij} = \sum_{k=1}^{r} G\left(\sum_{l=1}^{d} e_i(l+k) \wedge p_j(l) - s + 1 \right) \tag{5.8}$$

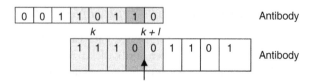

Figure 5.5 $p_j(l)$ is the *l*th element of string *p* representing antibody *j* (paratope *j*); $e_i(l + k)$ is the *l*th element of the *k*th shift of the bit string *e* representing antibody *i* (epitope *i*); $d = 8$.

where $p_j(l)$ denotes the *l*th element of the string *p* representing antibody *j* (paratope *j*); $e_i(l + k)$ denotes the *l*th element of the *k*th shift of the bit string *e* representing antibody *i* (epitope *i*) (see Figure 5.5). The Exclusive OR (XOR) operator is denoted by ^; in fact, the ^ operator is used to compute the Hamming distance between two binary strings; *d* is the minimum distance between the lengths of the two bit strings *e* and *p*; also, $r = d - s$. A threshold function is defined by *G*, $G(x) > 0$ if $x > 0$, and 0 otherwise. Note that the affinity defined in Equation 5.8 is not symmetric.

This model (Farmer et al., 1986) defines a set of differential equations for predicting the concentration of antibodies, based on the interaction between them and with foreign antigens. Let *N* be the number of different antibody types present in the IN; similarly, let *n* be the number of antigen types. Then the change in the concentration of antibodies of type *i* denoted by x_i is defined as

$$\frac{dx_i}{dt} = c\left[\sum_{j=1}^{N} m_{ji}x_i x_j - k_1\sum_{j=1}^{N} m_{ij}x_i x_j + \sum_{j=1}^{n} m_{ji}x_i y_j\right] - k_2 x_i, \quad \text{for } i = 1, ..., N \quad (5.9)$$

The first two terms model the interaction among antibodies; particularly, the first term represents the stimulation of the paratope of a type *i* antibody by the epitope of a type *j* antibody, whereas the second term represents the suppression of a type *i* antibody when its epitope is recognized by the paratope of a type *j* antibody.

In the third term, y_j represents the concentration of antigen of type *j*. The last term models the tendency of cells to die when no interaction with other antibodies occurs, and k_2 is the cell death rate.

An important aspect of this model is that the number of antigen and antibody types is considered dynamic. To update the total number of antigen and antibody types, a threshold on all their concentrations is defined. Thus the interaction of an antibody with all other antibodies and antigens is eliminated when the concentration drops below such threshold. The generation of new antibodies is done by applying genetic operators to the paratope and epitope strings using crossover, inversion, and point mutation. This process is closer to a model of genetic changes that occur during cloning than to the one that occurs when new types of antibodies are produced in the bone marrow.

Farmer et al. (1986) reported that the antibodies whose paratopes match epitopes are amplified at the expense of other antibodies. If the suppression and stimulation rate are the same (equal to 1 in their work) and $k_2 > 0$, then every antibody type will eventually die due to the damping term. However, if $k_1 < 1$, it favors the formation of reaction loops; thus, the numbers of loop can gain concentration, fighting the damping term. The number and lengths of the loops increase as N increases. Antibodies that do not recognize other elements are eventually discarded. Farmer et al. (1987) introduced an equation to describe the change in the concentration of antigen of type i:

$$\frac{dy_i}{dt} = k_4 \sum_{j=1}^{M} m_{ji} c_j y_i, \quad \text{for } i = 1, \ldots, n \tag{5.10}$$

Thus, Equation 5.10 describes the dynamics of intrinsic antigen elimination.

5.2.1.4 Parisi's Idiotypical Network

To study immunological memory, a simple IN model, which captures most of the qualitative features, was introduced by Parisi (1990). Parisi's model focuses on the behavior of the immune system in the absence of external antigens and attempts to find a global functional description of the IN.

Parisi's model assumes that auto-antibodies of a given antibody are a very large set of low responder clones and the connectivity of the IN is very high, and this network cannot be partitioned into subnetworks. This immune model seems to have similarity with the Hopfield model (Hopfield, 1982). Here, a fully connected network is considered, and a connection weight vector that represents the influence of antibodies on one another is defined. The concentration of any antibody, in the absence of external antigens, is considered to have only two values, either 0 or 1, and that the value is greater than 1 in the presence of stimulating antigen. The dynamics of the network in discrete time is further analyzed.

A matrix J_{ik}, which codes the effect of the kth antibody on the ith antibody is considered, similar to the synaptic weight matrix in a Hopfield network. The stimulatory effect of the network on the ith antibody is thus given by

$$h_i(t) = s + \sum_{j=1}^{N} J_{ij} x_j(t) \tag{5.11}$$

with $x_k(t) = \theta[h_i(t)]$, where $\theta(h)$ is a step function defined as 0 if h is negative, otherwise it will be 1. If J_{ik} is positive, then antibody k triggers the production of antibody i. In contrast, if J_{ik} is negative, then antibody k suppresses the production of antibody i.

The explanation of immunology memory is as follows: assuming that antibody Ab_1 is produced by modeling the stimulating antigen, the production of Ab_1 is increased in the presence of Ab_2. That is, the lifetime of Ab_1-producing cells is increased and so the population of Ab_1-specific helper T cells is also increased. Thus, Ab_2, considered as the image of the stimulating antigen, remains after the antigen is removed; thus, the presence of Ab_2 induces the survival of memory B cells.

5.2.1.5 Stewart and Carneiro's Idiotypical Network

In addition to the immune elements considered in the earlier models, this model introduces T cell cooperation (Stewart and Carneiro, 1999). Therefore, at a given time, the variables considered are

- z_i—the concentration of T lymphocyte clones of type i, for $i = 1, 2, ..., L$
- b_i—the concentration of antibody of B lymphocyte clones of type i, for $i = 1, 2, ..., M$
- f_i—the concentration of Ig molecules (free antibodies) they produce, for $i = 1, 2, ..., N$
- y_i—the effective concentration of antigens of type i, for $i = 1, 2, ..., R$

The following set of differential equations, which describe the dynamic behavior of the IN are defined.

$$\frac{dz_i}{dt} = -k_1 z_i + k_2 \alpha(\pi_i, \eta_i, z_i) + \xi_i, \quad \text{for } i = 1, 2, ..., L \tag{5.12}$$

where k_1 is the natural T cell death rate; $\alpha(\bullet)$ the amount of activated T cells, which is a function of both excitatory (π_i) and inhibitory signals (η_i) received by T clones; and $\xi_i(\bullet)$ the (thymic) production of T cell clones of type i.

Also, two sets of differential equations are introduced to describe the dynamic behavior of both bound antibodies and free antibodies. Accordingly, the change in antibodies bound to the surface of B cells follow the following differential equation:

$$\frac{db_i}{dt} = -k_3 \beta(\sigma_i, \tau_i, b_i) - k_4 b_i + \xi_i \tag{5.13}$$

where $\beta(\bullet)$ denotes the number of activated B cells in the clone, which is a function of the amount of both induction signals (σ_i) and the number of specific activated T lymphocytes available for cooperation (τ_i); k_4 is the natural B cell death rate; and $\xi_i(\bullet)$ represents the (thymic) production of T cell clones of type i.

Also, the change in the concentration of Ig molecules is defined as

$$\frac{df_i}{dt} = k_5\beta(\sigma_i, \tau_i, b_i) - (k_6 + k_7\sigma_i)f_i, \quad \text{for } i = 1, 2, \dots, N \tag{5.14}$$

where the first term represents the growth (at the rate k_5) of the amount of soluble Ig molecules as a proportion to the number of Ig-producing (activated) B cells (note that the function β is the same as in Equation 5.14); k_6 is the rate at which antibodies produced by clone i decreases; the factor $(k_7\sigma_i)$ is the change rate at which Ig molecules produced by clone i are removed due to the formation of complexes with available legends.

5.2.2 Discrete Immune Network Models

Different variations of discrete IN models have been studied; each model is distinguished by the abstractions of the network structure, its dynamics and metadynamics, and data representation.

5.2.2.1 Hunt and Cooke's Immune Network Model and Its Variation

The first AIN model (Hunt and Cooke, 1996) considers an IN of B cells that interact with one another according to their affinities. B cells are represented as binary strings, following some earlier works (Farmer et al., 1986); thus the affinity between B cells is defined based on the Hamming distance. If B cell stimulation by foreign antigens is above a certain threshold, then they will undergo cloning and mutation. Cloning produces a certain number of exact copies of a B cell. The number of copies, however, depends on the stimulation level of the B cell. Finally, in the simple substitution operator, a small (less than half) portion of the substring representing a B cell is replaced by the corresponding elements of another randomly selected B cell. Also, three types of mutation operators are introduced: multipoint mutation, substring regeneration, and simple substitution; however, at each time, only one of these operators is applied to a clone at random. In multipoint mutation, each element of the antibody is mutated with a certain probability. In substring regeneration, a substring of the antibody's paratope is selected at random to be replaced by a randomly generated string.

The training process is performed in an iterative fashion; at the end of each iteration, a proportion of the less-stimulated B cells are removed and replaced with newly generated B cells, which are incorporated in the network. A variation of this model is called artificial immune network (AINE), which was introduced by Timmis et al. (2000). In this model, B cells are represented by real-valued vectors, instead of binary strings. A similarity measure is then defined by the Euclidean distance between two B cells. The dynamics of the network is similar to Hunt and

Cooke's model. Also, the concept of a network affinity threshold (NAT) is intro-duced as a mechanism to control the density of the connections among antibodies of the IN; therefore, only the strongest connections between them are considered. In their model, affinity measure values are normalized to be in [0, 1]; then, if the affinity measure is less than the NAT, the original similarity measure is considered; otherwise, the lowest similarity value between two antibodies (i.e., 1) is assigned. Thus, a stimulation level (*sl*) is defined as

$$sl(x_i) = 1 + \left[\sum_{j=1}^{N} (1 - m(x_i, x_j)) - \sum_{j=1}^{N} m(x_i, x_j) + \sum_{j=1}^{n} m(x_i, y_j) \right] \quad (5.15)$$

if $sl(x_i) > \theta$, then $x_i = x_i + k(sl)$, else $x_i = x_i$.

The first two terms in Equation 5.15 represent the interaction of a type *i* anti-body with other antibodies. The first and second terms represent the inhibitory and excitatory signals from other antibodies, respectively. The last term represents the effect of antibody stimulation by foreign antigens. Here the NAT value is dynami-cally adjusted.

The AINE model was later modified to alleviate some difficulties such as popu-lation control and the calculation of the NAT. The modified model was called resource-limited AIN (RAIN), and the concept of artificial recognition ball (ARB), which is a representation of a number of identical B cells instead of a single B cell, was introduced. In this model, there is a resource pool (B cells) with centralized control and the ARBs compete for allocating such resources. Unlike AINE, in RAIN, those ARBs having zero resources are removed from the network, and the NAT is time-independent and derived from antigen dataset. The RAIN algorithm is presented in Figure 5.6.

Self-stabilizing artificial immune system (SSAIS). Neal (2002) proposed SAIS, which is based on RAIN for continuous analysis of time-varying data. Also, SSAIS does not consider B cell suppression while calculating the stimulation level.

Meta-stable memory IN. Neal (2003) proposed a modified version of SSAIS for data analysis, clustering, and immune memory. In this model, each ARB stimulation is done by foreign antigens and those neighbors, which are in a Euclidean space. Here, the cloning process is employed only during the pri-mary response, which is mediated by the NAT, but it does not consider muta-tion operator. In this model, ARBs having resources less than the defined mortality threshold are removed from the network.

5.2.2.2 Fractal Immune Network

This version (Bentley and Timmis, 2004) uses the concept of ARB and coined a new term "fractal recognition space" (FRS). Here, interactions among self-elements

algorithm is presented next.

RAIN( )
1. Initialization
Create an initial network, select the initial set of B cells as a subset of the antigens
2. Antigenic presentation
While not stopping criterion met *do*
for each antigen *do*
2.1. compute network stimulation levels and clonal selection
for each B cell *do*
compute stimulation level
2.2. metadynamics
Remove B cells with low stimulation level, via the resource allocation mechanism
2.3. clonal expansion
Select most stimulated B-cells and reproduce them in proportion to their stimulation level
2.4. somatic hypermutation
Mutate each clone inversely proportional to its stimulation level
2.5. network update
select mutated clones to be incorporated in the network

Figure 5.6   Details of the RAIN algorithm.

are done by artificial cytokines, which are represented by a single clone of the transmitting FRS. The signal is received by a fractal receptor (a clone of the receiving FRS) and then, the distance is calculated. If the distance is below a certain threshold and FRS is mature, then the transmitting FRS is stimulated. This stimulated FRS is cloned with a fixed probability and it is merged with the antigen by the merge process of fractal proteins. If there exists no such FRS, a new one is created at the antigen point as a primary response, like its parent model. At each iteration, based on the stimulation level, the FRS concentration is increased. If this concentration is below a mortality threshold, the element is removed from the network.

5.2.3   AiNet and Its Variations

This model is similar to RAIN and is proposed by De Castro and Von Zuben (2001). The main difference is that AiNet does not consider the stimulation concept, rather uses the affinity concept. Part of the network adaptation process is inspired by the clonal selection principle (Burnet, 1959). In AiNet, each network element corresponds to an antibody molecule. The affinity is used to remove redundant information from the network—if the affinity between two antibodies is greater than the suppression threshold, one of them is removed from the network. Figure 5.7 shows the AiNet algorithm.

The number of clones generated for one B cell (denoted as N_c) in the presence of an antigen is computed as

$$N_c = \sum_{i=1}^{N} round(N - d_{ij}N) \tag{5.16}$$

with N as the number of B cells in the population, d_{ij} the distance between the ith B cell and jth antigen, and $round(\cdot)$ is used to round a value to its closest integer.

5.2.3.1   Opt-aiNet

De Castro and Timmis (2002a,b,c) uses a version of AiNet for solving optimization problems, which is called opt-aiNet. This work assumes B cell clusters as an optimization problem (De Castro and Von Zuben, 2002a), where the center of a cluster corresponds to a local optimum of the fitness function (De Castro and Timmis, 2002a). Thus, clusters are expected to form around points with high fitness values.

In opt-aiNet (De Castro, 2003), B cells are encoded as real-valued vectors in a Euclidean space. Also, a fitness function to evaluate each B cell is defined based on an objective function to be optimized (either minimized or maximized). A population of B cells, considered as candidate solutions to the function being optimized, evolves in AiNet. Such B cells undergo a process of evaluation against the objective function, clonal expansion, mutation, selection, and evaluation of their affinities with other B cells in the population. Accordingly, opt-aiNet finds a B cell memory set that represents good values of the objective function. The network trains until it reaches a stable state, measured through the average fitness of the B cells (Figure 5.8).

If c is the cell to be mutated, then the resulting mutated cell c' (after affinity proportional mutation) is computed as

$$c' = c + \alpha\varepsilon \tag{5.17}$$

AiNet ()

1. Initialization

Create an initial network of antibodies

2. Antigenic presentation

While not stopping criterion met *do*

　　for each antigenic pattern *do*

　　　　　2.1. **compute affinities with current antigen**

　　　　　for each antibody *do*

　　　　　　compute affinity with current antigen

　　　　　2.2. **clonal selection**

　　　　　select a number of elements with high affinity and

　　　　　reproduce (clone) them proportionally to their affinity

　　　　　2.3. **metadynamics**

　　　　　remove memory clones whose affinity with current

　　　　　antigen is less than a predefined threshold

　　　　　2.4. **clonal interactions**

　　　　　determine the network interactions (affinity) among

　　　　　elements in the clonal memory set

　　　　　2.5. **clonal suppression**

　　　　　remove memory clones whose affinity with each other

　　　　　is less than a pre-specified threshold

　　　　　2.6. **network update**

　　　　　incorporate remaining clones of the clonal memory

　　　　　with all network antibodies

3. Compute antibody interactions

　　o　**network stimulations**
　　　　compute similarity between each pair of network antibodies

　　o　**network suppressions**
　　　　eliminate all network antibodies whose affinity is less

　　　　than a pre-specified threshold

　　o　**diversity handling**
　　　　introduce a set of new randomly generated antibodies

　　　　into the network

Figure 5.7　The AiNet algorithm.

---

original opt-aiNet()

1. Initialization

Create a random initial population of B cells (initial immune network)

2. Immune Network Dynamics

while not stopping criterion met *do*

    2.1. compute B cell fitness

    for each B cell *do*

        compute the fitness of current B cell

        normalize vector of all B cell fitnesses

    2.2. clonal expansion

    for each B cell *do*

        generate N_C clones of current B cell

        add new clones to current B cell population

    2.3. somatic mutation

    for each B cell clone *do*

        mutate each clone proportionally to the parent B cell's fitness

    2.4. fitness re-evaluation

    for each B cell *do*

        compute fitness of current B cell

    2.5. clonal selection

    select fittest clones and discard clones with the lowest fitness

    2.6. compute average fitness

    2.7. network supression

    suppress B cells whose affinities are below the suppression threshold σ_s

    2.8. memory cell differentiation

    determine memory cells after suppression.

    2.9. metadynamics

    introduce a percentage *d* of new randomly generated B cells into the network

---

Figure 5.8   The orginal opt-aiNet algorithm.

with ε (a Gaussian random variable with zero mean and standard deviation) equal to 1 (i.e., $\varepsilon \sim N(0,1)$) and $\alpha = (1/\varepsilon)e^{-f}$ is a factor that decays exponentially with the value of the B cell fitness f, which has been normalized to [0,1] and β is a parameter that controls the decay of the exponential function. In addition, a mutation can only be accepted if c' falls in the feasible space.

The opt-aiNet termination criterion is based on the size of the memory-cell population after network suppression. If the number of memory cells does not vary between subsequent network suppressions, then it is assumed that the network has reached stability and, therefore, the current population of memory cells gives a set of solutions of the problem at hand.

A feature of opt-aiNet is that it considers the interaction of the network cells with the environment (fitness) and with one another (affinity), allowing dynamical control of the size of the population. In opt-aiNet, new cells are allowed to enter the population only after the current cell population cannot significantly improve its average fitness. Opt-aiNet uses a Gaussian mutation that is inversely proportional to the normalized fitness of each parent cell. It also presents some general features similar to evolutionary strategies (ES). Selection mechanism is similar to a $(\mu + \lambda) - $ ES, in which a population of size μ parents generate λ offspring; the population formed by parents and offspring undergoes a selection process to reduce it to μ individuals again. Parents survive, unless they are suppressed by one of the offspring. In opt-aiNet, parameters μ and λ become N and N_c, the number of clones of each individual and size of the population, respectively.

Both opt-aiNet and ES use Gaussian mutation; however, opt-aiNet uses an affinity proportional to Gaussian mutation, whereas mutation used in ES is not based on fitness. Another important difference is that opt-aiNet allows variable population size, and the size of the population is dynamically adjusted through the introduction of diversity (network metadynamics) and discarding the least-fit B cells through network suppression. In contrast, the size of the population in ES is fixed.

Another model (IPD aiNet) was proposed by Alonso et al. (2004), which is a modification of aiNet model representing antigens and B cells as iterated prisoner's dilemma (IPD) strategies. The main modification is that if a B cell is added to memory, it will never be removed. The immune agent perceives the opponent's strategy and tries to find a strategy (most stimulated B cell), in the immune memory, which provides it the highest payoff to confront the playing opponent.

5.2.3.2 Dynamic Optimization AiNet

Olivetti et al. (2005) proposed a modification of opt-aiNet to deal with dynamic environment. Particularly, the following modifications are proposed:

■ The use of a separate memory subpopulation
■ A procedure to adjust the parameter that controls the decay of the inverse exponential function, denoted by β

- Introduction of two new mutation operator schemes
- A B cell linear suppression mechanism
- A limited population size

Dynamic optimization aiNet (dopt-aiNet) handles two separate populations: current- and memory subpopulation to control the growth of the population and avoid a decay in the performance due to an excessive growth of the population. The current subpopulation is the same as in opt-aiNet, new cells are incorporated into this subpopulation. However, the memory subpopulation represents local optima to which some cells converge. All B cells keep a rank value that is decremented whenever a mutation does not improve its corresponding fitness function value or otherwise, it gets incremented. If a cell's rank value reaches zero, then it becomes part of the pool of memory cells. B cells in memory population undergo special mutation operations as described as follows. After a B cell becomes a memory cell, it goes through a similar process (using a rank value) starting with a new rank value and after this new rank value gets to zero, a memory cell does not undergo any additional mutation.

A method called the golden section (Bazaraa et al., 1993) is used to find the best value of the parameter β, for each Gaussian random vector generated. This method divides an interval in two sections and determines which of the two is most promising and discards the other one. Such subinterval is then subdivided into two sections and the same process is subsequently repeated. This process continues until the remaining promising interval reaches a prespecified small length. An important parameter for this method is a ratio known as golden number or golden ratio (Bazaraa et al., 1993), which needs to be considered carefully while moving along the interval to find an optimum value of β; this method assumes that the function is continuous, convex, and unimodal, which is a very restrictive assumption for the fitness function, especially when dealing with real-world problems and given that opt-aiNet does not assume any additional information about the objective function.

To overcome this problem, the authors proposed a simple heuristics, which divides the interval into four subintervals, instead of only two. Then, the golden section method is applied to each subinterval separately, and the best result determines the current value of β.

Moreover, two new mutation operators introduced in dopt-aiNet are described in the following text. A one-dimensional mutation operator is used, which performs Gaussian mutation only in one dimension at a time. Supposedly, this will provide a finer search around a point. However, in high dimensions, this method presents slow convergence toward local optima. The other mutation operator introduced is called gene duplication, to emulate the process of gene duplication that sometimes occurs during chromosome transcription. Thus, a coordinate x_i is randomly chosen and copied into another coordinate x_j of the same B cell; and this change is taken if it improves the fitness of the B cell. In addition, the one-dimensional mutation operator is applied before applying gene duplication.

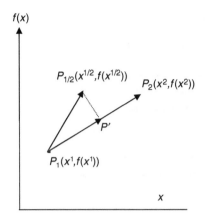

Figure 5.9   B cell suppression. Here *P'* is used to determine the suppression distance between the two points.

Also, a modification to the cell-suppression mechanism used in opt-aiNet is introduced. The new mechanism, termed "cell-line suppression," is proposed to reduce the probability of having more than one cell located at each peak of the fitness landscape. This suppression mechanism not only uses information of the domain space but also information of the fitness function as follows. When a B cell suppresses another B cell, instead of considering the distance between the points x^1 and x^2 that represent the two B cells, points of the form $(x, f(x))$ are used, which is described as follows. Let $P_1 = (x^1, f(x^1))$, $P_2 = (x^2, f(x^2))$, and $P' =$ projection of $P_{1/2}$ onto P_2, where $P_{1/2} = (P_1 + P_2)/2$ (Figure 5.9). The suppression between the B cells with values x^1 and x^2 is computed based on the distance between $P_{1/2}$ and a point P, which is computed as

$$P' = \begin{cases} P_1 + \left(\dfrac{v \cdot w}{\|v\|}\right)v & \text{if } P' \text{ falls inside segment } \overline{P_1 P_2} \\ P_1 & \text{if } P' \text{ falls outside segment } \overline{P_1 P_2} \text{ and is closer to } P_1 \\ P_2 & \text{if } P' \text{ falls outside segment } \overline{P_1 P_2} \text{ and is closer to } P_2 \end{cases} \quad (5.18)$$

where $v = P_2 - P_1$ and $w = P_{1/2} - P_1$. Accordingly, if $dist(P_{1/2}, P')$ is below a threshold value σ_s, then the B cell with the worst fitness between the two is removed.

To limit the growth of the population, a maximum number of cells is prespecified in such a way that when the B cell population reaches this value, B cells with the worst fitness are deleted from the population.

The dopt-aiNet algorithm is summarized in Figure 5.10.

Vargas et al. (2003) proposed the CLARINET model, which combines learning classifier systems, evolutionary algorithms, and AIN where classifier systems are

dopt-aiNet()

1. Initialization

Create a random initial population of B cells

(initial immune network)

2. Immune Network Dynamics

while not stopping criterion met *do*

   2.1. **compute B cell fitness**

   for each B cell *do*

   compute the fitness of current B cell

   normalize vector of all B cell fitnesses

   2.2. **clonal expansion and somatic mutation**

   for each B cell *do*

      generate N_c clones of current B cell

      add new clones to current B cell population

   for each B-cell *do*

      if mutated B-cell c' is better than original B cell c then

         $c.rank = c.rank+1$

         $c=c'$

      else

         $c.rank = c.rank-1$

      if $c.rank = 0$ then

         make c part of the B-cell memory pool

   for each B-cell *do*

      if $c.rank > 0$ *then*

         apply one-dimensional mutation to current B-cell

         apply gene-duplication mutation to mutated B-cell

         if mutated B-cell m is better than the original *then*

            $m.rank = m.rank+1$

         else

            $m.rank = m.rank-1$

   2.3. **compute average fitness**

   compute average fitness of the B-cell population

   2.4. **network supression and metadynamics**

   if average error stagnates apply B-cell suppression mechanism using

   suppression threshold σ_s and introduce a percentage d of new

   randomly generated B-cells into the network

   2.5. **population control**

   if average error stagnates and the number of B-cells exceeds

   the maximum number of B-cells

      then

      remove a percentage of less fit B-cells

Figure 5.10 The dopt-aiNet algorithm.

regarded as B cells that are interacting with one another through stimulation and suppression functions. Training algorithm uses same operations as defined in the earlier model.

5.2.3.3 Fuzzy Immune Network Model

Nasraoui et al. (2002) proposed an IN model that deals with uncertainty and fuzziness inherent in the matching process between antibodies and antigens, which is called a fuzzy artificial immune system (AIS), and is based on the AINE model of Timmis et al. (2002). In AINE, an IN consists of a set of interconnected ARBs, where each ARB is composed of identical B cells. Also, each ARB represents a single n-dimensional data item that may be stimulated by an antigen or another ARB. A link between two ARBs is created if the affinity between them is below a NAT, which is defined as the average distance between all the antigen items in the dataset provided as input to the training process. Antigen-ARB matching and the matching between ARBs are computed using an Euclidean distance. Also, when an ARB gets stimulated above the threshold, it undergoes some cloning and mutation processes. In addition, in AINE, the ARBs compete for a finite number of resources (B cells); thus, resources are allocated to ARBs as a function of their stimulation levels. Besides, ARBs that are not given any resources (i.e., have low stimulation level) are removed from the network. The purpose of the ARB concept is to reduce the granularity of the model. However, the ARB population grows at a prolific rate and it tends to converge rather prematurely to a network where a small number of internal images of the antigens overtake the entire population due to the resource allocation mechanism (Nasraoui et al., 2002).

In contrast to ARB, a fuzzy ARB does not represent a single data item but a fuzzy set over a universe of discourse defined by the input training data. The membership function associated to the fuzzy set is a radial-basis-type of function, which decreases with the distance to the point that represents the center of the ARB. In contrast to AINE, each fuzzy ARB is allowed to have its own radius of influence, denoted as σ. Particularly, the membership function associated to an ARB i is defined as

$$f_i(x) = e^{-\left(d(x,c_i)^2\right)/(2\sigma_i^2)} \tag{5.19}$$

where c_i is the center of the ARB, x an antigen, and $d(x,c_i)$ the distance from x to c_i. Thus, the stimulating level of ARB i produced by an antigen set X, denoted as $s_i(X)$, is computed as

$$s_i(X) = \frac{\sum_{x \in X} f_i(x)}{\sigma_2^i} \tag{5.20}$$

where X denotes an input antigen set, which generally consists of foreign antigens and other ARBs. Therefore, antigens closer to the center of the ARB will stimulate it more. In this fuzzy AIS (Figure 5.11), the value of the radius of influence of each ARB is also adapted at each iteration to maximize the ARB stimulation level and, hence, the ARB probability of survival. This is done by merging neighboring ARBs, when they are very close to other ARBs influence regions, which also help in limiting the growth of the population. Besides, some suppression effect among neighboring ARBs is introduced in the model. In this case, the influence of an ARB is adjusted by subtracting a suppression factor, due to neighboring antibodies, from the stimulation caused by the antigen set; and then this value is scaled accordingly. Thereby,

$$\sigma_i^2 = \frac{\sum_{x \in X} f_i(x) d^2(x, c_i) - \beta(t) \sum_{y \in A} f_i(y) d^2(y, c_i)}{\sum_{x \in X} f_i(x) - \beta(t) \sum_{y \in A} f_i(y)} \tag{5.21}$$

where A is the set of all the ARBs in the IN.

It is important to note that in the cloning process, the value of σ is inherited from the corresponding parent ARB. As mentioned earlier, the ARBs compete for a finite number of resources (B cells) and that the resources allocated depend on the fuzzy ARB stimulation levels. Therefore, to avoid this in the resource allocation process, the best ARBs overtake the whole population; the algorithm limits the influence of the best ARBs; and thus, the number of allocated resources r is computed in proportion to the total stimulation level sl as

$$r = k \cdot sl \tag{5.22}$$

where k is a constant.

In addition, to limit the growth of the population, if two fuzzy ARBs represent identical data (i.e., have identical centers) after cloning and mutation, then, they are merged into a single ARB. Accordingly, in fuzzy AINE, a postprocessing stage takes place to consolidate the final population of fuzzy ARBs as outlined in the pseudocode shown in Figure 5.12.

In this postprocessing process, crossover operation can be easily designed by randomly exchanging information of the centers of the two ARBs selected to be merged, or by computing the center of the new fuzzy ARB as a convex combination of the centers of the original fuzzy ARBs, if a real-vector representation is being used.

In experiments reported by Nasraoui et al. (2002), note that at the beginning of the evolution process, denser cluster dominate; but as the process goes on, those ARBs around less populated clusters start to increase their stimulation levels, until reaching the same stimulation levels as other ARB clusters. This is due to the way stimulated levels are computed today. Specifically, the penalization of neighbor fuzzy ARBs due to suppression, combined with the cloning mechanism that promotes the proliferation of highly stimulated fuzzy ARBs, prevents very good

Fuzzy AINE()

1. Initialization

Create an initial network, create initial set of ARBs with their particular σ_i based on the input antigen (data) set

2. Antigenic presentation

While not stopping criterion met *do*

    for each antigen *do*

            2.1. compute network stimulation levels and

                clonal selection

            for each fuzzy ARB *do*

                compute ARB stimulation level

                update σ_i

            2.2. metadynamics

            Allocate B cells to fuzzy ARB's based on stimulation level

            Remove fuzzy ARBs with low stimulation level

            2.3. clonal expansion

            Select most stimulated fuzzy ARBS and reproduce

            them in proportion to their stimulation level

            2.4. somatic hypermutation

            Mutate each fuzzy ARB inversely proportional to its

            stimulation level

            2.5. network update

            select mutated fuzzy ARBs to be incorporated in the

            network

Figure 5.11   The fuzzy AINE algorithm.

consolidating final fuzzy ARB population()
Antigenic presentation
for each pair of ARB's *do*
if affinity between two current ARB's is less than ε *then*
merge two ARB's in a single fuzzy ARB

Figure 5.12 A subfunction of fuzzy AINE.

fuzzy ARBs from dominating the least-stimulated ARBs, because their stimulation cannot go beyond certain limit. This process can be thought of as a niching mechanism to maintain diversity in the population of fuzzy ARBs.

5.2.3.3.1 The Dynamic Weight B Cell Model

Nasraoui et al. (2003b) proposed a modification of the fuzzy IN model to perform dynamic unsupervised learning. This model was proposed as an attempt to solve the scalability problems present in most IN problems, which in general need to manipulate a large number of antibodies (or B cells) and links that represent the interactions among them. This model is also intended to deal with dynamic environments. Thereby, antigens are presented to the IN one at a time. Accordingly, the stimulation levels and radius of influence of each ARB are updated after the presentation of each antigen.

In this model, the concept of a dynamic weighted B cell (D-W-B cell) is introduced. This is based on the concept of a fuzzy ARB, but instead of only considering a function to model the influence zone of an antigen, it also introduces a temporal aspect in the model. Then, the main assumption here is that more current data will have a higher influence on the network dynamics as compared to less current or older data. Thus, the membership function defined in Equation 5.22, after J antigens have been presented, becomes

$$f_i(x_j) = e^{-\left(\frac{d^2(x_j, c_i)}{2\sigma_i^2} + \frac{(J-j)}{\tau}\right)} \tag{5.23}$$

where $d(x, c_i)$ is the distance between the center of the ith D-W-B cell and antigen x_j, which is the jth antigen encountered by the IN. Accordingly, the stimulation level of the IN, after J antigens have been presented, is defined as

$$s_i(x_j) = \frac{\sum f_i(x_j)}{\sigma_i^2} \tag{5.24}$$

where x_j denotes the jth antigen presented to the IN. Hence, information about all the antigens that have been presented earlier to the network is used to compute the current stimulation level of each D-W-B cell. Also, corresponding equations to update the influence radii can be derived by making

$$\frac{\partial s_i(X)}{\partial \sigma_i^2} = 0$$

thus obtaining

$$s_i(x_j) = \frac{w_{ij} + e^{-\frac{1}{\tau}}W_{i,J-1}}{\sigma_{i,J}^2} \qquad (5.25)$$

where $W_{i,J-1} = \sum f_i(x_j)$, w_{ij} denotes $f_i(x_j)$, and $\sigma_{i,J}$ the radio of influence of ith D-W-B cell after presenting $(J-1)$ antigens, which is defined as

$$\sigma_{i,j}^2 = \frac{e^{\frac{1}{\tau}}\sigma_{i,J-1}^2 W_{i,J-1} + w_{ij}d_{i,J}^2}{2(e^{-\frac{1}{\tau}}W_{i,J-1} + w_{i,J})} \qquad (5.26)$$

where d_{ij} denotes the distance from D-W-B cell i to the Jth encountered antigen.

In this model, a dynamic stimulation factor $\alpha(t)$ is also incorporated in the calculation of the D-W-B cell stimulation level. This is done by allowing groups of D-W-B cells to have a stimulation coefficient that will cause the formation of sub-networks of D-W-B cell that can self-sustain, even after the antigen that caused their creation disappears from the environment. However, a limit should be specified on the time span that these patterns would contribute to the network dynamics so as to avoid the imposition of any additional superfluous (computational and storage) burden on the IN by such patterns. Thus, an annealing schedule is proposed for this stimulation factor, that is, this stimulation coefficient decreases with the age of the subnet.

In the absence of a recent antigen that succeeds in stimulating a given subnet, the age of the D-W-B cell increases to 1 with each antigen presented to the immune system. However, if a new antigen succeeds in stimulating a given subnet, then its age is reset to 0. Therefore, those subnets that are very old will gradually die, unless they become restimulated by recent relevant antigens. Incorporating a dynamic suppression factor in the computation of the D-W-B cell stimulation level is also a more reasonable way to take internal interactions into account. The suppression factor is not intended to control the proliferation and redundancy in the population of the D-W-B cells. It is important to note that the effect of positive stimulation is to provide a memory mechanism to the IN; however, suppression provides a mechanism to have the D-W-B cells compete to avoid redundancy.

This algorithm also proposes a hierarchical organization of the IN. Hence, the goal is to keep a small network organized in subnetworks. An interaction occurring between an external agent and any D-W-B cell in the IN is called an external interaction. However, those interactions occurring among D-W-B cells is called internal interactions. The main idea is then to cluster D-W-B cells to form subnetworks. Hence, if high-level (coarse) interactions are considered, then, inter-subnetwork interactions should be considered. Accordingly, in the case of external interactions, contacts between antigens and subnetworks are taken into account, instead of considering interactions of the antigen with all the D-W-B cells in the IN. This is intended to reduce the number of computations, because when external stimuli appear, only some subnetworks are stimulated. Figure 5.13 illustrates the hierarchical organization of the D-W-B cell network model and external and internal interactions. In contrast, if low-level relations are considered, then interactions among all the D-W-B cells in the network should be evaluated.

Clustering of D-W-B cells in subnetworks is done by associating a centroid to each cluster defined as the centroid of all the D-W-B cells in the cluster. Thus, the network is characterized by such centroid, which is used to compute the contributions of the D-W-B cells in internal and external interactions.

It is important to note that by clustering the D-W-B cells, a significant reduction on the computation may be achieved as follows. Assume that the network consists of N D-W-B cells. Then, the number of internal interactions is $O(N^2)$. For the sake of a general analysis, assume that the IN consists of K subnetworks of equal size. Thereby, the number of interactions of a D-W-B cell reduced to those interactions in the corresponding subnetwork (intra-network interactions) is $O(N^2/K)$; also, the number of inter-subnetwork interactions is $K - 1$.

5.3 A General Model of Artificial Immune Network

González et al. (2005) proposed an algorithm, which presents common features of various AIN models. This general AIN (GAIN) algorithm is shown in Figure 5.14. The first step of this algorithm creates an initial set of B cells denoted by B. The second step calculates the stimulation for each of the antigen and B cell. This can be represented as follows:

$$f^A_{stimulation} : A \times B \rightarrow \Re$$

In some models, the stimulation is the function of affinity, which is given as follows:

$$f^A_{stimulation}(a,b) := g(f_{affinity}(a,b))$$

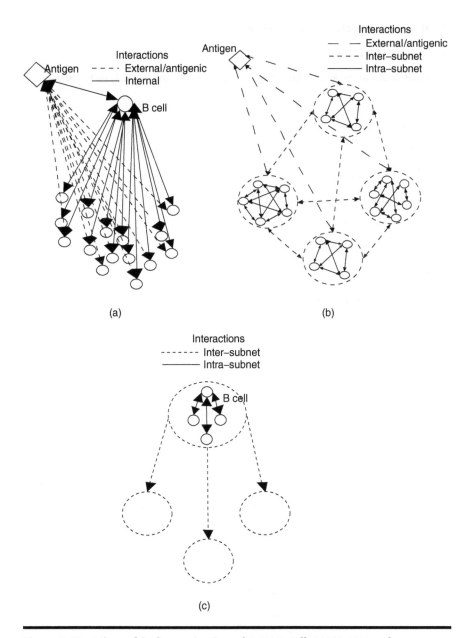

Figure 5.13 Hierarchical organization of D-W-B cell IN. (a) Network interactions without considering formation of subnetworks, (b) hierarchical organization of the IN, (c) internal IN interactions when a hierarchical structure is considered.

GAIN (General Artificial Immune Network)

Input: '*A*' as a set of antigens

Output: Immune network consisting of '*B*' as a set of B-cells and connections between them

1: Initialization

1.1: Assign *B* an initial set of B-cells

1.2: Initialize network structure *L*

2: **Repeat until** a stop criteria is met

2.1: Antigen presentation:

▷ Antigen/ B-cell affinity

2.1.1: Calculate $f_{affinity}(a,b) \; \forall \; a \in A, b \in B$

▷ Antigen/ B-cell stimulation

2.1.2: Calculate $f^A_{stimulation}(b,a) \; \forall a \in A, b \in B$

2.2: B-cell interaction:

▷ B-cell/B-cell stimulation/suppression

2.2.1: Calculate $f^B_{stimulation}(b',b)$ and $f^B_{suppression}(b',b) \forall b',b \in B$

2.3: Affinity maturation:

▷ Total stimulation

2.3.1: Calculate $F(b) := \sum_{a \in A, b' \in B, b' \neq b} f^A_{stimulation}(a,b) + f^B_{stimulation}(b',b) + f^B_{suppression}(b',b)$

2.3.2: Create $f_{cloning}(b)$ clones to the B-cell *b* and mutate then

2.3.3: Calculate stimulation of all new B-cells

2.4: Metadynamics:

▷ Detection/creation of B-cells and links

2.4.1: Update network structure *L*

▷ Return immune network

3: Return (B,L)

Figure 5.14 A GAIN algorithm. (From González, F., J. Galeano and A. Veloza, in *Proceedings of the 2005 Conference on Genetic and Evolutionary Computation (GECCO'05),* **ACM Press, Washington, 2005, 361–368.)**

where $f_{affinity}$ is the measure of the similarity and complementarity between elements in the shape–space and is given as $f_{affinity}$: B $\cup$ A $\times$ B $\cup$ A $\rightarrow$ R.

Now, the amount of stimulation produced by an antigen with a given affinity with the B cell is g: R $\rightarrow$ R.

Then, the 2.2 step calculates the B cell interaction by calculating the stimulation and suppression effects between them. These functions are given as follows:

$$f^B_{stimulation}\text{:B} \times \text{B} \rightarrow \Re$$

and

$$f^B_{suppression}\text{:B} \times \text{B} \rightarrow \Re$$

Then, the 2.2.1 step calculates the stimulation and suppression of B cell/B cell in the similar manner of antigen/B cell stimulation. Then, the total stimulation F: B $\rightarrow$ R is calculated by adding all the effects caused by the antigen and network connection, which is F(b) (given in step 2.3.1).

Now, some of the B cells are selected and $f_{cloning}(b)$ copies of each selected B cell, b are created. Now, mutation is done on these selected B cells. This mutation is varied from AIN model to model. After this, some of the B cells are deleted and some are again created randomly in the network and the connection or links between them are again created. Finally, this algorithm stops when the stopping criterion is met and returns the current network.

5.4 Summary

This chapter describes immune algorithms primarily based on clonal selection principle and idiotypic INs, particularly, the ClonAlg algorithm and its variations, which are based on the clonal selection and affinity maturation principles. The ClonAlg is similar to mutation-based evolutionary algorithms and has several interesting features: (i) dynamically adjustable population size, (ii) exploitation and exploration of the search space, (iii) location of multiple optima, (iv) capability of maintaining local optima solutions, and (v) defined stopping criterion.

The IN theory and continuous and discrete models are described. The derived computational algorithms are discussed in detail. Particularly, Hunt and Cooke (1996) proposed a supervised machine-learning algorithm based on IN model to classify DNA sequences as either promoter or nonpromoter classes. Timmis and Neal (2001) introduced another algorithm similar to it, but domain-independent, called AINE. This network constitutes a reduced version of the original data that can be used for data clustering or compression.

A major drawback of AINE is the explosion in B cell population. Thus, an enhanced algorithm called RAIN was developed (Timmis et al., 2000). The main difference between AINE and RAIN is that the basic element of the RAIN

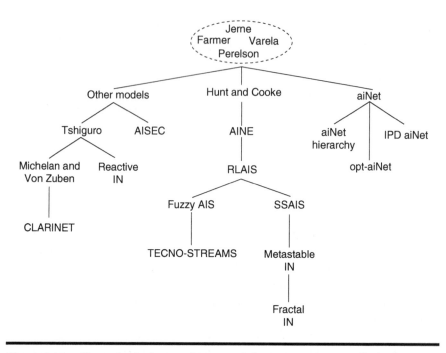

Figure 5.15   Chronological tree of AIN models. *RLAIS,* **Resource limited artificial immune system;** *AISEC,* **Artificial immune system for e-mail classification. (From González, F., J. Galeano, and A. Veloza in** *Proceedings of the 2005 Conference on Genetic and Evolutionary Computation (GECCO'05),* **ACM Press, Washington, 2005, 361–368.)**

algorithm is not the B cell but the ARB or cluster. Another model called aiNet, which shares some characteristics of Timmis' AINE, was proposed (De Castro and Von Zuben, 2001). This work emphasizes its self-organizing ability, that is, the use of minimal number of control parameters; several versions of AiNet with different enhancement are subsequently developed (De Castro and Timmis 2002a,b,c; De Castro and Von Zuben, 2002a,b; De Castro, 2003). Timmis and Edmonds (2004) have done further analysis of Opt-AiNet and commented on its implementation.

Nasraoui et al. (2002, 2003a) introduced an IN-based algorithm (called FuzzyAIS) that uses a fuzzy set to model the area of influence of each B cell. This improves the expression of earlier models and makes it more robust to noise and outliers.

In their recent work, González et al. (2005) developed a general model of AiNet (Figure 5.15), which provides a common notation and description of AiNet and its variation. Many applications of immune network models are reported in the literature (Ishiguro et al., 1994, 1996; Luh and Liu, 2004; Michelan and Zuben, 2002; Mitsumoto et al., 1996; Secker et al., 2003; Timmis and Edmonds, 2004; Timmis et al., 2004; Timmis and Neal, 2001).

5.5 Exercises

1. Find specific behavior for some of the IN models based on differential equations for some particular values of the parameters.
2. Design a specific learning rule for a parallel distributed processing (PDP) IN.
3. Use a specific learning algorithm to design a specific PDP IN.
4. Define IN basic elements such as shape–space and affinity measures for specific problems.
5. Ask some specific questions about particular differential equation models, such as under what conditions is stability reached? What is the behavior of the system in the absence of antigen? What are the attractors of the dynamical system? Also, answer specific questions such as the number of equilibrium points and limit cycle.
6. For AIN discrete models, also suggest some minor modifications to some of the models and find out how such changes modify the behavior of the network.
7. Design some specific AINs to solve toy problems such as
   a. Optimization of real value functions on 1D, 2D, or 3D
   b. Discrete optimization problems
   c. Small classification problems

5.6 General Questions

1. What is the general model of an IN, also including B cell–T cell cooperation?
2. How to model an AIN structure? (A straightforward way representing it as a graph in which nodes will correspond to antibodies)
3. How to characterize the different states of an antibody?
4. How to model an antigen?
5. How to specify the concept of an IN state?
6. How to model, in a general fashion, the process(es) that occur after the interaction with either antigens or antibodies (immune response) in such a way that the model conveys all different IN models? Describe the adaptation process that occurs?
7. Like neural networks, an IN dynamics should be characterized by a learning (adaptation) process that is typically a result of a stimulation process by a set of antigens. How does an antibody send signals to other antibodies as a result of the IN dynamics? How to express the total stimulation received by an antibody?
8. Explain different variations of clonal selection.
9. Distinguish between fuzzy AIS and a nonfuzzy version of AIN.
10. Describe the GAIN algorithm.

References

Alonso, O. M., F. Nino and M. Velez. A robust immune based approach to the iterated prisoner's dilemma. In G. Nicosia, V. Cutello, P. J. Bentley and J. Timmis (Eds.), *Proceeding of the Third Conference ICARIS*, Springer, Edinburgh, pp. 290–301, 2004.

Anderson, R. W., A. U. Neumann and A. S. Perelson. A Cayley tree immune network model with antibody dynamics. *Bull. Math. Biol.*, 55, 1091, 1973.

Bazaraa, M. S., H. D. Sherali and C. M. Shetty. *Nonlinear Programming: Theory and Algorithms*, 2nd Ed., Wiley, New York, 1993.

Bentley, P. J. and J. Timmis. A fractal immune network. In G. Nicosia, V. Cutello, P. J. Bentley and J. Timmis, (Eds.), *Proceeding of the Third Conference ICARIS*, Springer, Edinburgh, pp. 133–145, 2004.

Beyer, H.-G. and H.-P. Schwefel. Evolution strategies: A comprehensive introduction. *Nat. Comput.*, 1(1), 3–52, 2002.

Burnet, F. M. A modification of Jerne's theory of antibody production using the concept of clonal selection. *Aust. J. Sci.*, 20, 67–69, 1957.

Burnet, F. M. *The Clonal Selection Theory of Immunity*, Cambridge University Press, London, 1959.

Coutinho, A. A walk with Francisco Varela from first- to second generation networks: In search of the structure, dynamics and metadynamics. *Biol. Res.*, 36, 17–26, 1993.

De Boer, R. J. *Clonal Selection vs Idyiotypic Network Models of the Immune System: A Bioinformatic Approach*. PhD. Thesis, University of Utrecht, The Netherlands, 1989.

De Boer, R. J. and P. Hogeweg. Memory but no suppression in low dimensional symmetric idiotypic networks. *Bull. Math. Biol.*, 51, 223, 1989.

De Castro, L. N. The immune response of an artificial immune network (AiNet*). Proceedings of the IEEE Congress on Evolutionary Computation*, Canberra, December 8–12, 2003.

De Castro L. N. and J. Timmis. Artificial immune systems: A novel approach to pattern recognition. In L. Alonso, J. Corchado and C. Fyfe (Eds.), *Artificial Neural Networks in Pattern Recognition*, University of Paisley, Scotland, pp. 67–84, 2002a.

De Castro, L. N. and J. Timmis. An artificial immune network for multimodal function optimization. *Proceedings of the Special Sessions on Artificial Immune Systems in Congress on Evolutionary Computation, IEEE World Congress on Computational Intelligence*, Honolulu, HI, May 2002b.

De Castro, L. N. and J. Timmis. Convergence and hierarchy of aiNet: Basic ideas and preliminary results. *Proceedings of ICARIS (International Conference on Artificial Immune Systems)*, University of Kent, Canterbury, pp. 231–240, 2002c.

De Castro, L. N. and F. J. Von Zuben. The clonal selection algorithm with engineering applications. *Proceedings of Genetic and Evolutionary Computation Conference (GECCO) 2000*, Las Vegas, pp. 36–37, 2000.

De Castro, L. N. and F. J. Von Zuben. aiNet: An artificial immune network for data analysis. In H. A. Abbas, R. A. S. Charles and S. Newton (Eds.), *Data Mining: A Heuristic Approach*. Idea Group Publishing, Hershey, PA, pp. 231–259, 2001.

De Castro, L. N. and F. J. Von Zuben. opt-aiNet: An artificial immune network for multimodal function optimization. *Proceedings of IEEE Congress on Evolutionary Computation, CEC'02*, Honolulu, HI, vol. 1, pp. 699–674, 2002a.

De Castro, L. N. and F. J. Von Zuben. Learning and optimization using the clonal selection principle. *IEEE T. Evolut. Comput.* (special issue on artificial immune systems), 6(3), 239–251, 2002b.

Farmer, J. D., N. H. Packard and A. S. Perelson. The immune system, adaptation and machine learning. *Physica*, 22D, 187–204, 1986.

Farmer, J. D., S. A. Kauffman, N. H. Packard and A. S. Perelson. Adaptive dynamic networks as models for the immune system and autocatalytic sets. *Ann. NY Acad. Sci.*, 504, 118–131, 1987.

González, F., J. Galeano and A. Veloza. A comparative analysis of artificial immune network models. *Proceedings of the 2005 Conference on Genetic and Evolutionary Computation (GECCO'05)*, ACM Press, Washington, pp. 361–368, 2005.

Hopfield, J. Neural networks and physical systems with emerging collective computational abilities. *P. Natl. Acad. Sci.*, 79, 2554–2558, 1982.

Hunt, J. E. and D. E. Cooke. Learning using an artificial immune system. *J. Netw. Comput. Appl.* (special issue on Intelligent systems: Design and application), 19, 189–212, 1996.

Ishiguro, A., T. Kondo, Y. Watanabe, Y. Shirai and Y. Uchikawa. Immunoid: A robot with a decentralized consensus-making mechanism based on the immune system. *ICMAS Workshop on Immunity-Based Systems*, IEEE Computer Society Press, Washington, DC, pp. 82–92, 1996.

Ishiguro, A., S. Ichikawa and Y. Uchikawa. A gait acquisition of six-legged robot using immune networks. *Proceedings of International Conference on Intelligent Robotics and Systems (IROS '94)*, Munich, Germany, vol. 2, pp. 1034–1041, 1994.

Jerne, N. K. Towards a network theory of the immune system. *Ann. Immunol. (Paris)*, 125C, 373, 1974.

Luh, G.-C. and W.-W. Liu. Reactive immune network based mobile robot navigation. In G. Nicosia, V. Cutello, P. J. Bentley and J. Timmis (Eds.), *Proceeding of the Third Conference ICARIS*, Springer, Edinburgh, pp. 119–132 , 2004.

Michelan, R. and F. J. Von Zuben. Decentralized control system for autonomous navigation based on an evolved artificial immune network. *Proceedings of the IEEE Congress on Evolutionary Computation*, Honolulu, HI, vol. 2, pp. 1021–1026, 2002.

Mitsumoto, N., T. Fukuda, F. Arai, H. Tadashi and T. Idogaki. Self-organizing multiple robotic system. *Proceedings of the IEEE International Conference on Robotics and Automation*, Minneapolis, MN, pp. 1614–1619, 1996.

Nasraoui, O., C. Cardona, C. Rojas and F. González. TECNO-STREAMS: Tracking evolving clusters in noisy data streams with a scalable immune system learning model. *Third IEEE International Conference on Data Mining*, Melbourne, FL, 2003a.

Nasraoui, O., D. Dasgupta and F. Gonzalez. A novel artificial immune system approach to robust data mining. *Proceedings of the International Conference Genetic and Evolutionary Computation (GECCO)*, New York, July 9–13, 2002.

Nasraoui, O., F. González, C. Cardona, C. Rojas and D. Dasgupta. A scalable artificial immune system model for dynamic unsupervised learning. *Proceedings of the Genetic and Evolutionary Computation Conference (GECCO)*, LNCS 2723, Chicago, IL, 2003b.

Nasraoui, O., F. González and D. Dasgupta. The fuzzy artificial immune system: Motivations, basic concepts and application to clustering and web profiling. *IEEE International Conference on Fuzzy Systems*, Honolulu, HI, pp. 711–716, 2002.

Neal, M. An artificial immune system for continuous analysis of time-varying data. In J. Timmis and P. J. Bentley (Eds.), *Proceedings of the 1st International Conference on Artificial Immune Systems (ICARIS)*, University of Kent, Canterbury, vol. 1, pp. 76–85, 2002.

Neal, M. Meta-stable memory in an artificial immune network. In J. Timmis, P. Bentley and E. Hart (Eds.), *Proceedings of the Second International Conference ICARIS*, Springer, Edinburgh, pp. 168–180, 2003.

Neuman, A. U. *Dynamical Transitions and Percolation in Network Models of the Immune Response*. PhD. Thesis, Bar-Ilan University, Israel, 1992.

Olivetti de França, F., F. J. Von Zuben and L. N. De Castro. An artificial immune network for multimodal function optimization on dynamic environments. *Proceedings of the Genetic and Evolutionary Computation Conference (GECCO)*. Washington, pp. 289–296, 2005.

Parisi, G. A simple model for the immune network. *Proceedings of the National Academy of Sciences*, vol 87, pp. 429-433, 1990.

Secker, A., A. Freitas and J. Timmis. AISEC: An artificial immune system for e-mail classification. In R. Sarker, R. Reynolds, H. Abbass, T. Kay-Chen, R. McKay, D. Essam and T. Gedeon (Eds.), *Proceedings of the Congress on Evolutionary Computation*, IEEE, Canberra, pp. 131–139, 2003.

Stewart, J. and J. Carneiro. The central and the peripheral immune systems: What is the relationship? In D. Dasgupta (Ed.), *Artificial Immune Systems and their Applications*, Springer, Berlin, 1999.

Timmis, J. and C. Edmonds. *A Comment on Opt-AiNET: An Immune Network Algorithm for Optimisation*, GECCO, Seattle, Washington, pp. 308–317, 2004.

Timmis, J., C. Edmonds and J. Kelsey. Assessing the performance of two immune inspired algorithms and a hybrid genetic algorithm for function optimisation. *Proceedings of the Congress on Evolutionary Computation*, Portland, OR, vol. 1, pp. 1044–1051, 2004.

Timmis, J. and M. Neal. A resource limited artificial immune system for data analysis. *Knowl. Based Syst.*, 14, 121–130, 2001.

Timmis, J., M. Neal and J. Hunt. An artificial immune system for data analysis. *Biosystems*, 55, 143–150, 2000.

Varela, F. J. and A. Coutinho. Second generation immune networks. *Immunol. Today*, 12, 159–166, 1991.

Vargas, P. A., L. N. De Castro, R. Michelan and F. J. Von Zuben. An immune learning classifier system for autonomous navigation. In J. Timmis, P. Bentley and E. Hart (Eds.), *Proceedings of the Second International Conference ICARIS*, Springer, Edinburgh, vol. 2787, pp. 69–80, 2003.

Vertosick, F. T. and R. H. Kelly. Immune network theory: A role for parallel distributed processing. *Immunology*, 66, 1–7, 1989.

Weisbuch, G., R. J. De Boer and A. S. Perelson. Localized memories in idiotypic networks. *J. Theor. Biol.*, 146, 483, 1990.

Chapter 6

Latest Immune Models and Hybrid Approaches

This chapter covers some immune-computing concepts, which have not been well developed yet but appear to have potential in providing more insights of immune processes. One such model, which has gained a lot of attention recently is the so-called danger theory (DT) (Matzinger, 1994, 2002). Thus, this chapter introduces the main ideas behind DT and briefly describes some computational models based on the DT inspired by dendritic cells (DCs) functionalities. Other recent works presented include a multilevel immune algorithm, major histocompatibility complex (MHC)-based approaches, and cytokine networks. The last section discusses a hybrid approach, which combines negative selection (NS) and neural network methods to design a classification algorithm.

6.1   The Danger Theory

The DT (Matzinger, 1994) states that the immune system is activated on receipt of molecular signals, which indicate damage (or stress) to the body rather than by pattern matching of "nonself" versus "self." Accordingly, distressed cells and tissues transmit danger signals, which results in capturing antigens by antigen-presenting cells (APCs) such as macrophages; APCs then travel to the local lymph node and present the antigen to lymphocytes. Essentially, a danger zone exists around each danger signal. Therefore, only those B cells whose antibodies match antigens in the danger zone will get stimulated and then will undergo clonal expansion. Figure 6.1 illustrates immune response described by the DT.

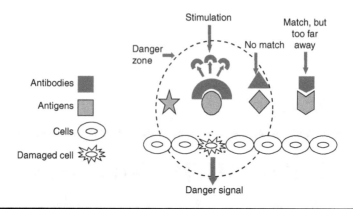

Figure 6.1   Illustration of the DT (the figure indicates the danger zone, danger signal, etc.).

The danger model can be seen as an extension of the two-signal model proposed by Bretscher and Cohn (1970). Two signals considered in this model are *antigen recognition* and *costimulation*. In DT, a costimulation signal indicates that an antigen is dangerous. To understand lymphocyte behavior, DT introduces the following three laws of lymphocytes (Matzinger, 1994):

- *Law 1.* A lymphocyte becomes activated if and only if it receives both signal 1 and signal 2. In the absence of one of these signals, lymphocytes remain inactive.
- *Law 2.* A lymphocyte accepts signal 2 only from APCs (or, in the case of B cell, it gets from T helper (T_h) cells). B cells can act as APCs only for experienced (memory) T cells. It is also important to note that signal 1 may come from cells other than just APCs.
- *Law 3.* Once activated, lymphocytes do not need signal 2; they revert to a resting state after a short period of time.

These rules usually apply to all mature lymphocytes, so immature cells are unable to accept signal 2 from any source. Thus, some screening by the NS process occurs first. Also, activated (effector) cells respond only to signal 1, ignoring signal 2, and revert to the resting state shortly afterward. Figure 6.2 illustrates different models using immune signals and progression in the development of DT.

In Burnet's (1959) original immune model, only signal 1 was considered, which is shown in Figure 6.2a. In this model, the only signaling occurs between infectious agents and lymphocytes (B cells and T killer cell, T_k). Consideration of a second signal was introduced by Bretscher and Cohn (1970) as shown in Figure 6.2b; here signal 2 comes from a T_h cell after receiving signal 1 from a B cell. Particularly, B cell presents an antigen to a T_h cell and waits for a confirmation signal. If the T_h cell recognizes the antigen (as nonself or dangerous), then an immune response is triggered. Another work (Lafferty and Cunningham, 1975) proposed that the T_h cell itself needs to be activated by signal 1 and signal 2 coming from APCs (see Figure 6.2c). It is important to note that T_h cells receive signal 1 from two sources: B cells and APCs.

The next model introduced by Janeway (1992) used the notion of INS, such as bacteria, which "primes" APCs causing the production of signal 2; in Figure 6.2d, priming signal is labeled as signal 0.

However, Matzinger (1994) proposes that the priming of APCs is due to a danger signal from stressed tissues or cells (Figure 6.2e). She also proposed extending the efficacy of T_h cells by routing signal 2 through APCs. This signal is marked as "signal 3" in Figure 6.2f. According to her, the antigen seen by the T_k cell does not need to be the same as the T_h; the only requirement is that both must be presented by the same APC. This allows T_h cells to prime many more T_k cells than they would otherwise have been able to. The DT can be seen as a natural extension of immune signal models proposed earlier. Figure 6.3 exhibits the partitioning of the antigen universe based on three models: self/nonself (SNS), INS, and DT (Matzinger, 2002).

According to Matzinger (2007), there is a category of damage-associated molecular patterns (DAMPs) that encompasses both pathogen-associated molecular patterns (PAMPs) and alarm signals. The ultimate control lies with the tissues in which the response occurs, rather than with the pathogen against which it is directed. Particularly, tissues use all sorts of mechanisms to keep the cells and molecules triggering immune responses in order to control the invaders. Many of these cells also seem to recognize stress-induced "self" molecules rather than foreign pathogens.

According to this view, at least some immune responses are initiated by tissue-derived signals that activate and educate APCs to control the effector class of an immune response. Accordingly, some danger signals such as tissue damage trigger a myriad of immune reactions and responses.

One of the problems of DT, however, is that the exact nature of danger signals is unclear. Also, there are some danger signals that should not trigger an immune response such as cuts or transplants. In addition, DT is not able to explain autoimmune diseases.

6.1.1 Danger Theory–Based Algorithms

Aickelin and Cayzer (2002) include the following aspects of the DT in their artificial immune system (AIS) design principles:

- Appropriate APCs to present danger signals need to be modeled.
- A danger signal can be either positive or negative, which means the presence or absence of the signal.
- Although in biology the danger zone is spatial, in computation model other notions of proximity, such as temporal proximity, may be used.

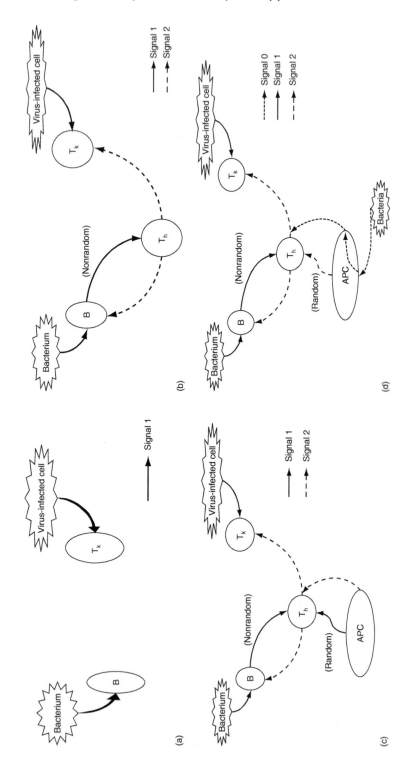

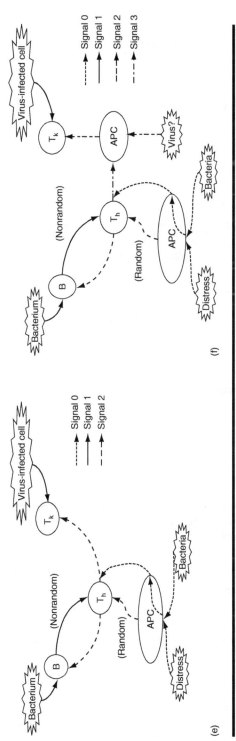

Figure 6.2 Shows the progression of different models based on the interaction of immune cells through various signaling. (From Aickelin U. and S. Cayzer. *Proceedings of the First International Conference on Artificial Immune Systems,* 2002.) (a) One signal model; signal 1 is between infectious agents and lymphocytes. (From Burnet F., *The Clonal Selection Theory of Acquired Immunity,* Vanderbilt University Press, Nashville, TN, 1959.) (b) Two signals model; signal 2 comes from T$_h$ after receiving *signal 1* from B cell. (From Bretscher, P. and M. Cohn, *Science,* 169, 1042–1049, 1970.) (c) This model introduced APC in controlling immune response. (From Lafferty K. and A. Cunningham, *Aust. J. Exp. Biol. Medical Sci., 53,* 27–42, 1975.) (d) The concept of infectious nonself (INS) is proposed with a third signal. (From Janeway, C.A., *Immunol. Today* 13, 11–16, 1992.) (e) The danger in control through zoning. (From Matzinger, P., *Annu. Rev. Immunol.,* 12, 991–1045, 1994.) (f) Multiplication of effect through interaction. (From Matzinger, P., *Science,* 296, 301, 2002.)

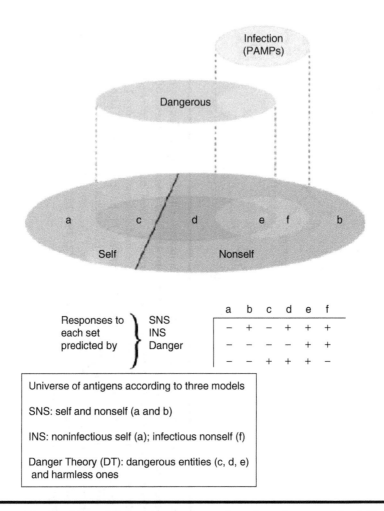

Figure 6.3 Partition of the antigen universe based on three models: SNS, INS, and DT. (From Matzinger, P., *Science*, 296, 301, 2002.)

- In biology, sometimes killer cells cause a normal cell death; however, this should not generate further danger signals.
- Models should appropriately consider priming killer cells via APCs for greater effect.
- Antibody migration rules should specify how many antibodies receive signal 1 and signal 2 from a given APC. DT relies on concentrations of different kinds of immune cells.

These aspects of DT are being used to build better AISs for anomaly detection. In this case, nonself patterns do not trigger an immune response without a danger signal

(Aickelin et al., 2003). The authors subsequently developed two algorithms—the dendritic cell algorithm (DCA) (Greensmith et al., 2006) and the toll-like receptor (TLR) algorithm (Twycross, 2007). These algorithms focus on different aspects of innate immunity to develop the basis of computation models.

6.1.2 Combining Dendritic Cells and Danger Theory

Yeom (2007) used a similar approach of combining the DT and DCs to form a schema for the signal precategorization. This categorization is based on the following principles:

- PAMPs are proteins expressed exclusively by bacteria, which can be detected by DCs indicating an anomalous situation.
- Danger signals are produced as a result of unplanned necrotic cell death. On damage to a cell, the chaotic breakdown of internal components forms danger signals, which accumulate in tissue. DCs are sensitive to changes in danger signal concentration. The presence of danger signals may or may not indicate an anomalous situation; however, the probability of an anomaly is higher than under normal circumstances.
- Safe signals are produced through the process of normal cell death. Cells must die for regulatory reasons, and the tightly controlled process results in the release of various signals into the tissue. These "safe signals" result in immune suppression. The presence of safe signals almost certainly indicates that no anomalies are present.
- Inflammatory cytokines can be released as a result of injury, although the process of inflammation is not enough to stimulate DCs alone, it can amplify the effects of the other three categories of signal.

DCs can stimulate naive T cells and have a number of functional properties (Yeom, 2007).

- DCs' first function is to instruct the immune system to act when the body is under attack, policing the tissue for potential sources of damage.
- DCs perform different functions based on their state of maturation. Modulation between these states is facilitated by the detection of signals within the tissue, namely, danger signals, PAMPs, apoptotic signals (safe signals), and inflammatory cytokines.
- In tissue, DCs collect antigen (regardless of the source) and experience danger signals from necrosing cells and safe signals from apoptotic cells. Maturation of DCs occurs in response to the receipt of these signals.

According to Yeom (2007), if there is a concentration of danger signals in the tissue at the time of antigen collection, the DCs become fully mature DCs with license (LmatDCs), and express LmatDCs cytokines. Conversely, if the DC is exposed to safe signals, the cell matures differently and becomes an unlicensed mature DC, expressing ULmatDCs cytokines. The LmatDCs cytokines activate T lymphocytes expressing complimentary receptors to the presented antigen. Any peripheral cells expressing this antigen type are removed through the activated T lymphocyte. The ULmatDCs cytokines suppress the activity of any matching T cell, inducing tolerance to the presented antigen.

6.2 Multilevel Immune Learning Algorithm

A multilevel framework that combines both B- and T cell recognition mechanisms was proposed, and called the Multilevel Immune Learning Algorithm (MILA), which is inspired by the interaction and processes of T cell–dependent humoral immune response (Dasgupta et al., 2003). In biological immune systems, some B cells recognize antigens through immunoglobulin receptors on their surface, but they are unable to proliferate and differentiate unless prompted by the action of lymphokines secreted by T_h cells.

Moreover, for T_h cells to become stimulated to release lymphokines, they must also recognize specific antigens. However, although T_h cells recognize antigens through their receptors, they can only do so in the context of MHC molecules. Antigenic (Ag) peptides are extracted by APCs through a process similar to feature extraction, called *Ag presentation*. Under certain conditions, however, B cell activation is suppressed by T suppressor (T_s) cells, but specific mechanisms for such a suppression are still unknown. The activated B and T cells migrate to the primary follicle of the cortex in lymph nodes, where a complex interaction and kinetic process of proliferation (cloning), mutation, selection, differentiation, and death of B cells take place in germinal center chambers. These antibodies function as effectors to the humoral response by binding to antigens and facilitating their elimination.

In MILA, an abstraction of these complex immunological events is incorporated to develop a multilevel change detection algorithm. Accordingly, the algorithm consists of initialization, recognition, evolutionary, and response (as shown in Figure 6.4).

In the *initialization* phase, the detection system is "trained" by giving the knowledge of "self." The outcome of the initialization is used to generate sets of detectors, analogous to the populations of T_h-, T_s-, and B cells, which participate in humoral response. It is important to note that a multilevel (multiresolution) detection is considered; specifically, three levels of detection were introduced:

- APC level, which corresponds to the highest level
- B cell level, the intermediate level, is used for global pattern
- T_h cell, the lowest level, or bit level, for local patterns

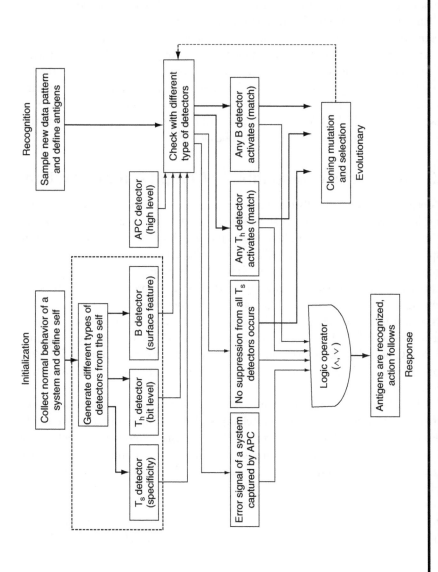

Figure 6.4. Overall description of MILA, showing different phases and detection strategies.

The operators used to combine different pattern recognition mechanisms can be chosen in different ways either to achieve more fault-tolerant recognition or sensitivity to small changes. MILA used an rcb-like matching rule in a real-valued representation, which may be thought of as a partial euclidean distance. A T_h cell uses a slide window to get the w elements. However, a B cell uses randomly chosen w elements. The concept of permutation mask and crossover closure from string presentation can be used the way these w elements are chosen for B cells.

Another feature of MILA is the implementation of positive selection by the so-called T_s cells (positive detectors), which are based on self-samples. An evolutionary phase in MILA is a process of refining the detector set if the earlier detection rates can be evaluated (verified). This phase involves cloning, mutation, and selection; however, cloning in MILA is a targeted one (not blind), only those detectors that are activated in the recognition phase can be cloned.

6.3 Major Histocompatibility Complex–Based Systems

To detect virus-infected cells that have been damaged internally, MHC molecules provide a mechanism to see what happened inside a cell because under normal circumstances, one cell cannot look inside another. The primary role of MHC is to display antigen fragments to T cells, particularly, MHC class I binds to CD8 on cytotoxic (or killer) T cells, whereas MHC class-II proteins present antigen fragments on an APC surface for binding to the CD4 receptor on the T_h cells. By allowing T cells to examine the internal state of other cells, the MHC mechanism acts as a kind of cell-level anomaly detector that allows the immune system to uncover virus-infected cells. A similar analogy of cells with running programs, and MHC peptides with short sequences of system calls was first proposed in 1996 (Forrest et al., 1996), which could detect malicious "running" programs such as viruses and worms.

All biological systems maintain a stable internal state by monitoring and responding to internal and external changes. This self-monitoring is one of the defining properties of life and is known as "homeostasis." Homeostatic mechanisms are usually autonomic, and the purpose is to minimize variations in the internal state of an organism. Somayaji and Forrest (2000) developed process homeostasis (pH), a Linux kernel extension that delays the execution of unusually behaving processes. The delay mechanism of pH was a predecessor to the network-level virus throttling (Twycross and Williamson, 2003), where the rate of network connections was throttled to limit the spread of malicious programs (Somayazi, 2007).

6.4 Cytokine Network Model

Various immune cells in the biological immune system mutually influence one another's activities through hormonelike intercellular messenger molecules called cytokines. The cytokines produced by one cell can modulate the production and secretion of cytokines by other immune cells. These interactions form a lymphoid endocrine system called the cytokine network. The immune cells in the cytokine network perform a dual role, producing cytokines and carrying out the immune response (acting as effector cells). The cytokines produced by the network regulate the development and growth of the responding immune cells.

From a computational point of view, the input in the cytokine network has the state of disease (extent and severity) and the state of the ongoing responses (extent and efficacy), whereas the output includes the proliferation of selected effector cells and the organization of new responses.

The model of the cytokine network proposed by Horne and van den Bergt (2007) considered an intercellular medium in which n distinct chemical species of cytokines diffuse and are well mixed. Both the input and output are encoded by the concentrations of the various cytokines. The artificial cytokine network includes the following elements:

- Cytokine concentrations
- Cytokine types
- External stimuli
- The density of cell for each type

Cytokine production by a cell of any one of these types depends on external stimuli as well as the cytokines themselves. The system as a whole is functional, mapping the external stimuli into a cytokine profile. The kinetics of the cytokine network is divided into two types: the dynamics of the cytokines and the dynamics of the cytokine-producing cells. The following equation was used to calculate the dynamics of the cytokines:

$$\dot{u}_k = \sum_{l=1}^{m} \psi_{lk}(u_1, \ldots, u_n, s_1, \ldots, s_r) \upsilon_l - \upsilon_k u_k$$

where u is the cytokine concentration and s is the external stimuli. Function $\psi_{lk} > 0$ expresses the effect of the cytokines and external stimuli on the production of cytokine k by a cell of type l; $\upsilon_k > 0$ is the rate of degradation of the kth cytokine.

The dynamics of the cytokine-producing cells was measured by the following equation:

$$\dot{\upsilon}_l = (\varphi_l(u_1, \ldots, u_n, s_1, \ldots, s_r) - \mu_l)\upsilon_l$$

where $\varphi_l > 0$ expresses the effect of the cytokines and external stimuli on the proliferation rate of a cell of type l and $\mu_l > 0$ is the death rate of cells of type l.

The results from the simulation with only two cytokines and one type of cell show that the stronger second stimulus produces a stronger response than the first, in terms of the increased concentration of cytokine-producing cells (Horne and van den Bergt, 2007).

6.4.1 Phylogenies of T Cells

Cytokines play an important role in T cell–mediated cellular immunity, which kills virus-infected cells and tumor cells. Different populations of T cells are regulated through signal transduction antigen-receptor-mediated pathways (Figure 6.5).

- T-delayed hypersensitivity (T_{DH}) lymphocyte, also known as the Th1, CD4+ T cell produces lymphokines to direct the cell-mediated immune response. It produces IL2 (T cell growth factor) that all T lymphocytes must have to respond to antigen.
- T_h lymphocyte (T_H), also called Th2, CD4+ T cell helps stimulate the B cell response.
- T cytotoxic lymphocyte (T_{cy}) is a CD8+ lymphocyte that kills vital infected cells and tumor cells.
- T memory cells (T_m) remember immune response, which are of CD4+ or CD8+ types depending on the memory.

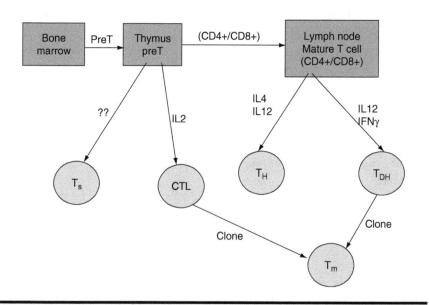

Figure 6.5   Cytokine network in T cell phylogenesis.

- T_s cells suppress humoral-immune response by suppressing the T_H. If T_{DH} cell is suppressed, then there is no cellular immune response. If T_{cy} is suppressed, then tumor and viral immunity is affected.

In innate immune response, some cytokines such as IL2, IL4, IL12, and IFNγ are produced by macrophages; these cytokines can determine the generation of different types of T cells.

Although there is no computational model yet published on T-cell phylogenesis, this provides another area of further research in immunological computation.

6.5 Combining Negative Selection and Classification Techniques

In many anomaly detection applications, only positive (normal) samples are available at the training stage. However, most conventional classification algorithms need both self- and nonself samples.

To allow conventional algorithms to be used when only samples from one class are provided, a hybrid algorithm was proposed (González et al., 2002), which is able to create synthetic nonself samples from a set of self-samples. The algorithm generates a set of detectors (points) that cover the nonself space using NS, and then these points are used as samples of the nonself class, allowing the use of conventional classification algorithms (Figure 6.6).

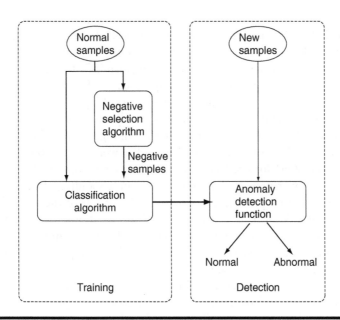

Figure 6.6   Combining NS-SOM in generating classifier dataset.

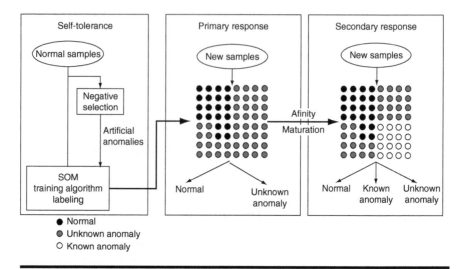

Figure 6.7   NS-SOM model structure. The model consists of three phases: self-tolerization, primary response (affinity maturation), and secondary response. The squared arrangement of nodes corresponds to an SOM, where black, gray, and white labels represent normal, unknown anomaly, and known anomaly respectively. (From González, F. J., Galeano, A. Veloza and A. Rojas. *Proceedings of the 2005 Conference on Genetic and Evolutionary Computation (GECCO'05)*, ACM Press, New York, 2005.)

Particularly, this approach uses the positive (normal) samples to generate negative samples. Then samples from both the classes are used (as training data) for a neural network, particularly, for a self-organizing map (SOM). Figure 6.7 shows the algorithmic steps of this hybrid NS-SOM approach. An SOM, composed of nodes or neurons (that are able to recognize different types of inputs), is a type of artificial immune network that is trained to produce a low-dimensional representation of the input space or self-/nonself feature space of the training samples called "map."

González et al. (2005) extended their earlier works of combining NS algorithm and the SOM for producing the visual representation of the self-/nonself feature space. This representation provides the understanding of the structure of self-/nonself space by producing a visual discrimination of the normal, known abnormal, and unknown abnormal regions. This model produces a network that can discriminate normal samples from abnormal samples and can learn from the encounters with antigens to improve specificity of response. This NS-SOM anomaly model has three phases: self-tolerization, primary response, and secondary response. These three phases are illustrated in Figure 6.7.

The first phase, self-tolerization, uses NS algorithm to produce the artificial anomalies and then, these anomalies are used to produce an anomaly classifier by using an SOM training algorithm instead of using classifier training algorithm.

This SOM training algorithm produces a network where nodes in SOM represent the structure of normal and abnormal samples for training, and nodes are labeled on the basis of this representation. Most importantly, the first phase uses NS algorithm for generating normal space and this normal space is used to train an SOM to produce a map that only reflects the self-space leaving the nonself space.

During the second phase, primary response, if unlabeled nodes are present, the network can classify them as normal or abnormal on the basis of finding the node, which is present near the input (the winner code) where the inputs are classified on the basis of the label of this node. Node labels are assigned by finding the closest labeled sample for each node and assigning the sample's labels for the corresponding node. This strategy would generalize k-nearest neighbor classification.

In the third phase, secondary response, new unlabeled nodes are classified on the basis of winner node. In this phase, the network is supposed to work more precisely by producing a specific response and by identifying more accurately about the normal or specific kind of anomaly.

The first stage is executed only once, but the second and third phases are repeated until there are new samples available. A visual representation of the feature space can be generated by a two-dimensional (2-D) grid corresponding to the network and by assigning different colors to each node depending on the category it represents. González et al. (2005) experimentally showed that this model can capture the structure of the normal samples used for training and by the third phase, it was seen that this model incredibly improved the discrimination by the affinity maturation process that is carried on the second phase of the model.

6.6 Summary

This chapter first discusses the DT, then DT- and DC-based methods are described. Other recent works presented include a multilevel immune algorithm, MHC-based approaches, and cytokine networks. Recent works in immunology show that as the antigen evolve toward imitating self-molecules, antigens became invisible to the antibodies' defense mechanisms pointing to the necessity of other means of protection probably constituted by T cells (Dasgupta, 2007). Moreover, the phylogenies of immune system (Warr and Cohen, 1991), the evolutionary development of immune functions play an important role in keeping diversity. The last section discusses a hybrid approach, which combines NS and neural network methods to design a classification algorithm.

6.7 Review Questions

1. Explain the main ideas behind DT.
2. How can a danger zone be defined?
3. Illustrate with diagrams self-nonself (SNS), infection/noninfectious (INS), and danger signal models.

4. Why is a danger model seen as an extension of the two-signal model?
5. Explain lymphocyte behavior in the DT.
6. Based on the laws of lymphocytes, explain why immature cells are unable to accept signal 2 from any source?
7. Briefly explain why the DT can be seen as an extension of immune signal models.
8. Draw flow diagram to explain the role of danger signal in the BIS.
9. Why is the DT not able to explain autoimmune diseases? Illustrate with an example.
10. What are the unique features of dentritic cell–based algorithm?
11. How are MHC-based and pH methods used for computer security?
12. What are the phases of MILA? State the T cell types that are being used in MILA.
13. Explain cytokine network model; identify its uniqueness and usefulness in real world applications.
14. Why do most conventional classification algorithms need both self- and non-self samples?
15. Briefly explain the idea behind NS-SOM (González et al., 2005) algorithm to create synthetic nonself samples from a set of self-samples.

References

Aickelin, U., P. Bentley, S. Cayzer, J. Kim and J. McLeod. Danger theory: The link between ais and ids? *Proceedings of the Second International Conference on Artificial Immune Systems (ICARIS 2003)*, vol. 2787 of LNCS, Springer-Verlag; pp. 147–155, 2003.

Aickelin, U. and S. Cayzer. The danger theory and its application to artificial immune systems. *1st International Conference on Artificial Immune Systems*, Canterbury, U.K., 2002.

Aickelin, U. and J. Greensmith. Sensing danger: Innate immunology for intrusion detection. *Inf. Security Tech. Rep.*, 12(4), 218–227, 2007.

Bretscher, P. and M. Cohn. A theory of self-nonself discrimination. *Science*, 169, 1042–1049, 1970.

Burnet, F. *The Clonal Selection Theory of Acquired Immunity*, Vanderbilt University Press, Nashville, TN, 1959.

Dasgupta, D. Immuno-inspired autonomic system for cyber defense. *Inf. Security Tech. Rep.*, 12(4), 235–241, 2007.

Dasgupta, D., S. Yu and N. S. Majumdar. MILA—Multilevel Immune Learning Algorithm. *Proceedings of Genetic and Evolutionary Computation Conference (GECCO)*, Illinois, Chicago, pp. 183–194, July 2003.

Forrest, S., S. A. Hofmeyr, A. Somayaji and T. A. Longstaff. A sense of self for Unix processes. *Proceedings of the IEEE Symposium on Security and Privacy*, IEEE Computer Society Press, Los Alamitos, CA, pp. 120–128, 1996.

González, F., J. Galeano, A. Veloza and A. Rojas. Discriminating and visualizing anomalies using negative selection and self-organizing maps. *Proceedings of the 2005 Conference on Genetic and Evolutionary Computation (GECCO'05)*, ACM Press, Washington, pp. 297–304, June 25–29, 2005.

González, F., D. Dasgupta and R. Kozma. Combining negative selection and classification techniques for anomaly detection. *Proceedings of IEEE Congress on Evolutionary Computation*. Honolulu, Hawaii, vol. 1, pp. 705–710, May 2002.

Greensmith, J., U. Aickelin and J. Twycross. Articulation and clarification of the dendritic cell algorithm. *ICARIS-06*, LNCS 4163, Oeiras, Portugal, pp. 404–417, 2006.

Hofmeyr, S., A. Somayaji and S. Forrest. Intrusion detection using sequences of system calls. *J. Comput. Security*, 6, 151–180, 1998.

Hone, A. and H. van den Bergt. Modelling a cytokine network. *Proceedings of the IEEE Symposium on Foundations of Computational Intelligence (FOCI)*, Honolulu, HI, 2007.

Janeway, C. A. The immune system evolved to discriminated infectious nonself from non-infectious self. *Immunol. Today*, 13, 11–16, 1992.

Lafferty, K. and A. Cunningham. A new analysis of allogeneic interactions. *Aust. J. Exp. Biol. Med. Sci.*, 53, 27–42, 1975.

Matzinger, P. Tolerance, danger and the extended family. *Annu. Rev. Immunol.*, 12, 991–1045, 1994.

Matzinger, P. The danger model: A renewed sense of self. *Science*, 296, 301, 2002.

Matzinger, P. Friendly and dangerous signals: Is the tissue in control? *Nat. Immunol.*, 8(1), 11–13, 2007.

Somayaji, A. Immunology, diversity, and homeostasis: The past and future of biologically-inspired computer defenses. *Inf. Security Tech. Rep.*, 12(4), 228–234, 2007.

Somayaji, A. and S. Forrest. Automated response using system-call delays. *The Proceedings of the 9th USENIX Security Symposium*, Denver, Colorado, August 14–17, 2000.

Twycross, J. Integrated innate and adaptive artificial immune systems applied to process anomaly detection. PhD thesis, School of Computer Science, The University of Nottingham, 2007.

Twycross, J. and M. M. Williamson. Implementing and testing a virus throttle. *Proceedings of the 12th USENIX Security Symposium*, Washington, D. C., pp. 285–294, 2003.

Warr, G. W. and N. Cohen. *Phylogenesis of Immune Functions*, CRC Press, Boca Raton, FL, 1991.

Yeom, K.-W. Immune-inspired algorithm for anomaly detection. *Stud. Comput. Intell. (SCI)*, 57, 129–154, 2007.

Chapter 7

Real-World Applications

Immunological computation (IC) techniques (or artificial immune systems) have been used as a problem solver in a wide range of domains such as optimization, classification, clustering, anomaly detection, machine learning, adaptive control, and associative memories. They have also been used in conjunction with other methods (hybridized) such as genetic algorithms (GAs), neural networks, fuzzy logic, and swarm intelligence. IC includes real-world applications of computer security, fraud detection, robotics, fault detection, data mining, text mining, image and pattern recognition, bioinformatics, games, scheduling, etc.

First, a general description of the solution process of using immune-based models is presented followed by some general-purpose applications. Next, some applications of AISs are briefly described to exhibit how these techniques can be used in real-world problem solving.

7.1 Solving Problems Using Immunological Computation

To apply an immunity-based model to solve a particular problem in a specific domain, one should select the immune algorithm depending on the type of problem that needs to be solved. Accordingly, the first step should be to identify the elements involved in the problem and how they can be modeled as entities in a particular AIS. To encode such entities, a representation scheme for these elements should be chosen, such as a string representation, real-valued vector, or hybrid representation. Subsequently, appropriate affinity/distance measures, which are to be used to determine corresponding matching rules, should be defined. The next step

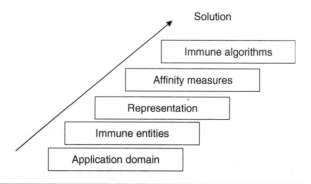

Figure 7.1 Solving a problem using IC.

should be to decide which AIS will be better to generate a set of suitable entities that can provide a good solution to the problem at hand. Figure 7.1 shows the necessary steps to solve problems using an immunological approach.

7.2 Applications in Computer Security

Computer security seems to be analogous to the biological defense in many respects (Forrest et al., 1996, 1997); thus we can learn a lesson from the immune system to develop digital immunity. Majority of AIS works have been devoted to using some immunological metaphor for developing digital defense systems (Aicklen et al., 2003; Dasgupta, 1999; D'haeseleer et al., 1996). AISs used varied notions of data protection and anomaly to provide a general-purpose protection system to augment current computer security systems. The security of computer systems depends on activities such as detecting unauthorized use of computer facilities, maintaining the integrity of data files, and preventing the spread of computer viruses. This immunity-based system is much more sophisticated.

Forrest et al. (1994) first proposed the usage of negative selection (NS) in computer security. They assumed the problem of protecting computer systems from harmful viruses as an instance of the general problem of distinguishing "self" (legitimate users, uncorrupted data, etc.) from the dangerous "other" (unauthorized users, viruses, and other malicious agents). This method was intended to be complementary to the more traditional cryptographic and deterministic file-authentication methods on the problem of computer virus detection.

7.2.1 Virus Detection

In this application, the NS algorithm (discussed in Section 4.2) was used to detect changes in the protected data and program files. Experiments were performed in a Disk Operating System (DOS) environment with different viruses, including file-infector and boot-sector virus samples, and the reported results showed that the method could detect modifications in the data files due to virus infection. When compared to other virus detection methods, this algorithm exhibits several advantages over the existing change detection methods: it is probabilistic and tunable (the probability of detection can be traded off against central processing unit [CPU] time), it can be distributed (providing high system-wide reliability at low individual cost), and it can detect novel viruses that have not been identified previously (Forrest et al., 1994).

However, because the stored information in a computer system is volatile in nature, the definition of self in computer systems should be more dynamic than the biological notion of self. For example, computer users routinely load in updated software systems, edit files, or run new programs. Therefore, this implementation seems to have limited use (only to protect static data files or software).

7.2.2 An Alternative Approach to Virus Detection

Kephart (1994) proposed a different immunologically inspired approach (based on instruction hypothesis) for virus detection. In this approach, known viruses are detected by their computer-code sequences (signatures) and unknown viruses by their unusual behavior within the computer system. This virus detection system continually scans a computer's software for typical signs of viral infection. These signs trigger the release of "decoy programs" whose sole purpose is to become infected by the virus.

Specifically, a diverse suit of decoy programs are kept at different strategic areas in the memory (e.g., home directory) to capture samples of viruses. According to the author, decoys are designed to be as attractive as possible to trap those types of viruses that spread most successfully. Each of the decoy programs is examined from time to time to see if it has been modified. If one or more have been modified, it is almost certain that an unknown virus is loose in the system, and each of the modified decoys contains a sample of that virus. Particularly, the infected decoys are processed by "the signature extractor" to develop a recognizer for the virus. It also extracts information from the infected decoys about how the virus attaches to its host program (attachment pattern of the virus), so that infected hosts can be repaired. The signature extractor must select a virus signature (from among the byte sequence produced by the attachment derivation step) such that it can avoid both false-negatives and false-positives while in use. In other words, the signature must be found in each instance of the virus, and it must be very unlikely to be found in uninfected programs. Once the best possible signature is selected from candidate signatures of the virus, it runs against a half-gigabyte corpus of legitimate programs to make sure that they do not cause a false-positive. The repair information is checked by testing on samples of the virus, and further by a human expert.

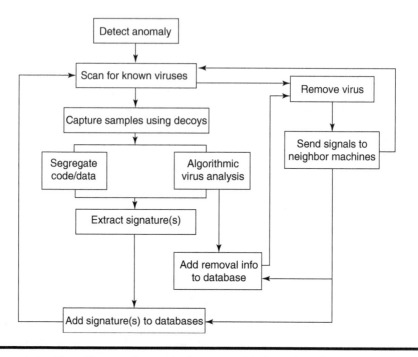

Figure 7.2   Flow diagram shows Kephart's approach in virus detection.

Finally, the signature and the repair program are stored in an archive of the anti-virus database, and the updated (new) version needs to be distributed to the customers (Figure 7.2).

In another work, Lamont et al. (1999) proposed a computer virus immune system (CVIS), which uses hierarchical intelligent agent architecture for identifying, attacking, and eradicating viruses from computers and networks. Particularly, coordination among intelligent agents is accomplished at three levels: local, network, and global. The functions of agents at each level are shown in Figure 7.3. For example, an agent at local level monitors an individual computer (or node) for potential viruses, where each node uses decoy program as described earlier (Kephart, 1994). Agents at the network level keep track of viruses in network traffic and inform at the local level, whereas agents at the global level involve in generating and adapting virus-fighting resources.

An automated detection and response system for identifying malicious self-propagating code and to stop its spread, called Cooperative Automated worm Response and Detection ImmuNe Algorithm (CARDINAL), was proposed by Kim et al. (2005). This method was based on the concepts of differentiation states of T cells. Particularly, three key properties of T cells have been identified: T cell proliferation to optimize the number of peer hosts polled, T cell differentiation to

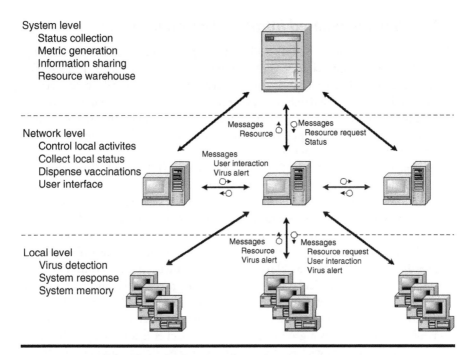

System level
  Status collection
  Metric generation
  Information sharing
  Resource warehouse

Network level
  Control local activites
  Collect local status
  Dispense vaccinations
  User interface

Messages
Resource

Messages
Resource request
Status

Messages
User interaction
Virus alert

Local level
  Virus detection
  System response
  System memory

Messages
Resource
Virus alert

Messages
Resource request
User interaction
Virus alert

**Figure 7.3   Multilevel model for virus detection. (From Lamont, G. B.,
R. E. Marmelstein and D. A. Van Veldhvizen.** *New Ideas in Optimization*
(edited volume), McGraw-Hill, 1999, 167–184.)

assess attack severity and certainty, and T cell modulation and interaction to balance local and peer information. The goal of this work was to use diverse T cell
types to operate as a cooperative automated worm detection and response system.

7.2.3   UNIX Process Monitoring

Forrest et al. (1996) applied an NS algorithm to monitor UNIX processes in a
dynamic computer environment in such a way that the definition of self is sensitive to malicious attacks. This work is based on the assumption that the system
calls of root processes are inherently more dangerous in causing damage than user
processes. Also, root processes have a limited range of behavior, and their behavior is relatively stable over time; accordingly, the "normal (or self)" is defined by
short-range correlations in a process' system calls (called trace). They experimented
with several common intrusions involving "sendmail," such as traces of successful
sendmail attacks, traces of sendmail intrusion attempts that failed, and traces of
error conditions.

7.2.4 Immunity-Based Intrusion Detection Systems

Further works by Hofmeyr et al. (1998) in computer security led to the development of host-based intrusion detection systems, which construct a database that catalogs the normal behavior over time in terms of the system calls made, etc. As this record builds up, the database may be monitored for any system calls that are not found in normal behavior patterns. Hofmeyr et al. argued that while simplistic, this approach is not computationally expensive and has the advantage of being platform and software independent.

Hofmeyr and Forrest (1999, 2000), Somayaji et al. (1998), and Warrender et al. (1999) conducted extensive research on an artificial immune system called ARTIS architecture, which could tackle the issue of protecting networks of computers. This is achieved in a similar way in monitoring network services, traffic and user behavior, and attempts to detect misuse or intrusion by observing departures from normal behavior. Each computer runs a broadcaster, which broadcasts the source and destination of each TCP SYN packet it sees, to other computers running LISYS (a version of ARTIS). Particularly, a detection node processes the information from the broadcasters. Each detection node receives data from broadcasters and mails it to the administrator if it detects a novel TCP connection. A detection node has an array of detectors that as a group determine if a packet is anomalous. Detectors are randomly generated, with each one sensitive to a particular random source and destination address, and port as well as near matches to it. For a newly generated detector, if it sees a packet that matches its template, a new randomly generated detector will replace it. For a detector over a week old, if it recognizes a packet, it will send a mail to the administrator for inspection. By having this weeklong "tolerization" period for the new detectors, they can generate detectors randomly and only let the ones that do not send false-positives for a week "survive." When the user receives an alarm signal from a detector, if the user does nothing, the detector that flagged the connection as anomalous will disappear and not bother the user any more. If the user chooses to confirm or "costimulate" the anomaly, the detector that flagged the anomaly will become a permanent part of the program's repertoire and will alert the user whenever this TCP connection is being requested in the future (shown in Figure 7.4).

Balthrop et al. (2002) used a version of LISYS for monitoring network traffic. The system used an NS algorithm (to mature 49-bit binary detectors, that is, triplets representing Transmission Control Protocol [TCP] connections), which was tested against connections collected during a training period. Matured detectors were then distributed in each host on a live network (see Figure 7.5). Diversity was created through each host independently reacting to its self and nonself (normal and abnormal). The matching function used was r-contiguous, and the detectors were improved through affinity maturation. It used a distributed detection strategy wherein each detection node, through a different representation filters incoming strings through a randomly generated permutation mask. This technique of having

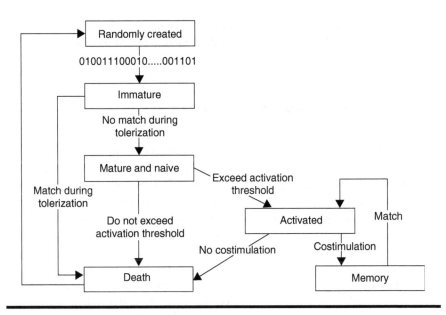

Figure 7.4   Shows the flow diagram of detector tolerization process.

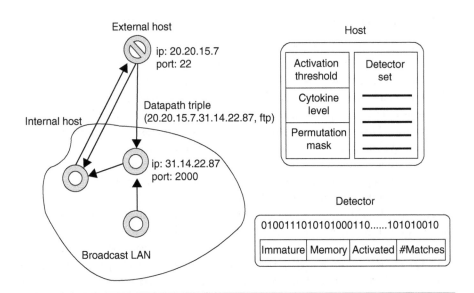

Figure 7.5   Illustrates the distributed detection scheme, where each host contains different detector sets. Each detector is a binary triplet representing TCP connections.

different representations for each detection node is equivalent to multiple detector shapes (hence changing the shape of the detectors), while keeping the "shape" of the self-set constant.

7.2.5   Immune Agent Architecture

An immune agent architecture was introduced by Dasgupta (1999) where immunity-based agents roamed in the machines (nodes or routers) and monitored the situation in the network (i.e., looked for changes such as malfunctions, faults, abnormalities, misuse, deviations, and intrusions). These agents could mutually recognize each other's activities and took appropriate actions according to the underlying security policies. Specifically, their activities were coordinated in a hierarchical fashion while sensing, communicating, and generating responses. These agents simultaneously monitored the networked computer's activities at different levels (such as user level, system level, process level, and packet level) to make robust decision on intrusions or anomalies. Some agents used B cell mechanism, some used T cell, and some had limited life cycle (time-dependent functionalities). Such architecture appears to be flexible and extendible, where an agent can learn and adapt to its environment dynamically and can detect both known and unknown intrusions.

7.2.6   Immunogenetic Approaches in Intrusion Detection

Gonzalez (2002) proposed negative selection with detector rules (NSDR) to detect attacks by monitoring network traffic. A real-valued representation was used for evolving hyper-rectangular-shaped detectors, interpreted as "if-then rules," for high-level characterization of the self/nonself space (i.e., normal and abnormal traffic). Experiments were performed using the 1999 Defense Advanced Research Projects Agency (DARPA) intrusion detection evaluation dataset. This data represents normal and abnormal information collected in a test network, in which simulated attacks were performed. The immunogenetic approach was able to produce detectors that gave a good estimation of the amount of deviation from the normal.

Further works extended the NSDR algorithm to use fuzzy detection rules, and is called NSFDR. This improves the accuracy of the method and produces a measure of deviation from the normal that does not need a discrete division of the nonself space. It provides a better definition of the boundary between normal and abnormal. The earlier approach used a discrete division of the nonself space, whereas the new approach does not need such a division because the fuzzy character of the rules provides a natural estimate of the amount of deviation from the normal. It shows an improved accuracy in the anomaly detection.

In another work, Kim and Bentley (2001) used three evolutionary stages: gene library evolution, negative selection, and clonal selection with the goal of designing effective network intrusion detection systems. Here, detectors (in the form of classifiers) are evolved using the clonal selection algorithm wherein the evolving population of detectors is clustered into "niches," which help to distinguish between self and nonself in network traffic data.

7.2.7 Danger Theory in Network Security

Aicklen et al. (2003) first proposed to use the danger theory (DT) concept in intrusion detection. Their system behaves like the dendritic cells (DC) looking for danger signals such as sudden increases in network traffic or unusually high numbers of error messages. If these signals increase above a preset threshold, it triggers an alert. Subsequently, two algorithms were developed based on the DT, the dendritic cell algorithm (DCA; Greensmith et al., 2006), and the Toll-like Receptor algorithm (TLR; Twycross, 2007). These algorithms focus on different aspects of innate immunity to develop the AIS models; a brief description is provided in the following.

7.2.7.1 Dendritic Cell Algorithm

It is an abstraction of DC functions, which is based on the premise that "suspects" in the form of antigen can be paired with "evidence" in the form of signals to identify potential sources of anomaly or intrusion. A general overview of the DCA is provided by Greensmith et al. (2006). The DCA is implemented using the libtissue framework to facilitate the creation and updating of cells and tissue attributes. A schematic diagram of the DCA is presented in Figure 7.6. The algorithm processes two input streams consisting of signals and antigens (data to be correlated). Particularly, the signal stream contains a specified number of input signals, which are prenormalized and categorized as pathogen-associated molecular pattern (PAMP), danger signal, safe signal, or inflammation. A storage facility for incoming signals and antigen is provided and forms the "tissue" for the DCs. The DCA can be described on two levels: first, at the level of an individual DC and second, at the level of the DC population. Similar to the biological immune system, DCs exist in one of the following three states—immature, semimature, or mature.

7.2.7.2 TLR Algorithm

Algorithmic steps of TLR algorithm (as described in Aickelin and Greensmith, 2007), which is primarily designed or anomaly detection in computer networks are provided as follows:

1. Record set of system calls (low-level instructions in computing) made in training data.
2. Record signal values experienced in training data.

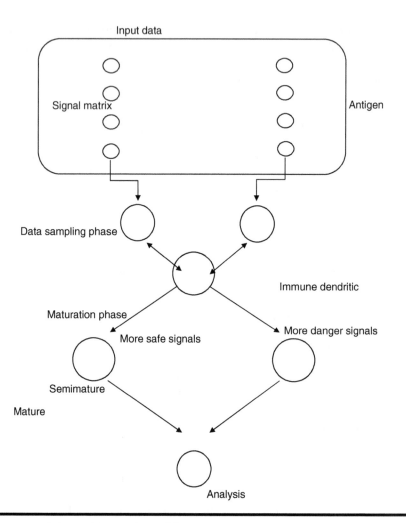

Figure 7.6 Schematic overview of the DCA.

3. Compute complement set to 1 and 2.
4. Create immature DCs (iDCs) with signal receptors randomly drawn from the complement signal set and with antigen receptors randomly drawn from the complement system call set.
5. Create naive T cells (nTCs) with antigen receptors randomly drawn from complement system call set.
6. iDCs are continually exposed to sample signals and antigens, respectively.
7. If during its lifetime an iDC's signal receptor matches a signal, it becomes a mature DC (mDC) and migrates.
8. If an iDC has not migrated at the end of its lifetime, it becomes a semimature DC (smDC) at the end of its lifetime and migrates.

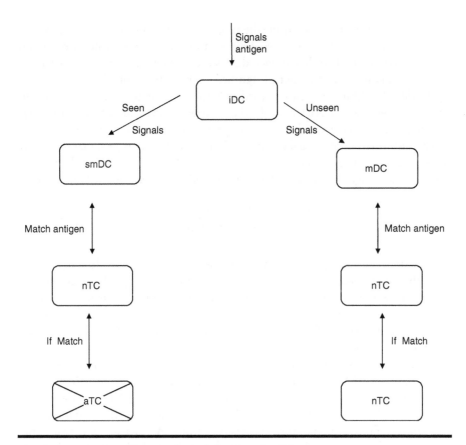

Figure 7.7 Systematic overview of the TLR algorithm.

9. Migrated smDCs and mDCs present their antigen and try and match nTCs.
10. If an mDC expresses an antigen that matches an nTC receptor, then this turns the nTC into an activated T cell (aTC) and we have an anomaly.
11. If an smDC expresses an antigen that matches an nTC receptor, then this kills the nTC to reduce false-positives.
12. Migrated smDCs and mDCs and killed nTCs are replaced with new cells as per points 4 and 5 (Figure 7.7).

There are some similarities between the DCA and TLR algorithms; both perform a type of temporal correlation between signals and an antigen. However, TLR defines interactions between both T cells and DCs, which is more complex than the single cell and multiple-signal model employed by the DCA.

Yeom (2007) extended a DT-based approach for network anomaly detection, where input signals are combined with some information such as data length, name or identification (ID), or process/program service ID. Here, data is combined with context information received during the antigen collection process. Different combinations of input signals result in two different antigen contexts: "unlicensed mature antigen" context implied that antigen data was collected under normal conditions, whereas a "mature antigen" context signified a potentially anomalous data.

In this algorithm, antigen was used only for the labeling and tracking of data and hence, represented as a string of either integers or characters. Signals were represented as real-valued numbers that are proportional to values derived from the context information of the dataset in use. For example, a danger signal may be an increase in CPU usage of a computer. The value for the CPU load can be normalized within a range and converted into its real-valued signal concentration value.

$$C_{[csm,LmatDCs,ULmatDCs]} = \frac{((9W_P * C_P) + (W_S * C_S) + (W_D * C_D))}{W_P + W_S + W_D} * \frac{(1 + IC)}{2} \quad (7.1)$$

In Equation 7.1, the signal values are combined using a weighted function, where Cx is the input concentration and Wx is the weight. Input signals are categorized either as PAMPs (P), safe signals (S), danger signals (D), or inflammatory cytokines (IC), and are represented as a concentration of signals. They are transformed to output concentrations of costimulatory molecules (csm), ULmatDCs cytokines, and LmatDCs cytokines.

To detect port scan attacks, three different signals—PAMPs, danger, and safe—are used, where PAMPs indicated the number of "unreachable destination" errors. When the port scan process scans multiple Internet Protocol (IP) addresses indiscriminately, the number of these errors increases. Danger signals are indicative of the number of outbound network packets per second. An increase in network traffic could imply anomalous behavior. The safe signals are the inverse rate of change of network packets per second. This is based on the assumption that if the rate of sending network packets is highly variable, the machine is behaving suspiciously (Figure 7.8 shows the pseudocode for this DC approach).

7.3 Applications in Fraud Detection

An immune-based system called JISYS was applied to fraud detection (Hunt and Cooke, 1995; Hunt and Fellows, 1996; Hunt et al., 1999). This system forms a network of B cell objects where each B cell represented a loan application. Advances were made in the follow-on work by Hunt et al. (1999) where results for fraud detection were presented. Work presented by Neal et al. (1998) discusses an immune-inspired supervised learning system called "Immunos-81." Two standard machine-learning datasets were used to test the system's recognition capabilities. They use software abstractions of T cells, B cells, antibodies, and their interactions. Artificial T cells control the creation of B cell populations (clones), which compete for recognition of "unknowns." The B cell clone with "simple highest avidity" (SHA) or "relative highest avidity" (RHA) is considered to have successfully classified the unknown.

```
PROCEDURE InitializationOfStatus()
    While (size of AB < a threshold)
        If (detector expresses NOSYMTOMS in network)
            create new antibody (ab) from log file
            add newly generated antibody (ab) to AB
            For (ab ∈ AB)
                clone and mutate ab to maximize affinity with ab
                add best n clones to AB

PROCEDURE Continuous_adaptation()
    antigen_count ← 0
    LOOP
        receive incoming network packets
        ag ← preprocess packets into antigen
        antigen_count ← antigen_count +1
        IF (antigen_count = K)
            AG ← last K antigens
            Update_Population(AG)
            antigen_count ← 0
        compute degree of danger()
        WHILE(danger is high)
            compute temporal danger zone
            AG ← all packets in the danger zone
        FOREACH (ag ∈ AG)
            FOREACH (ab ∈ AB)
                compute affinity (ab,ag)
            high_aff ← highest affinity value
            IF (high_aff > a threshold)
                block the scanned port
```

Figure 7.8   The pseudocode for detecting scan probes.

Research by Carter (2000) attempted to create an inductive computation algorithm based on metaphors taken from immunology. They describe an evolutionary search algorithm based on the model of immune network dynamics.

7.4   Application in Robotics and Control

Robot control works by Ishiguro et al. (1996, 1998), Wantanabe et al. (1998, 1999), and Lee et al. (1999) focused on the development of a dynamic decentralized consensus-making mechanism based on the "immune network theory." They attempted to create a mechanism by which a single, self-sufficient autonomous robot, called the immunoid, could perform the task of collecting various amounts of garbage from a constantly changing environment. The authors used the metaphors of antibodies, which were potential behaviors of the immunoid; antigens corresponded to environmental inputs such as existence of garbage, wall, and home bases (Figure 7.9). For the immunoid to make the best decision, it detects antigens and matches the content of the antigen with a selection of all the antibodies that it possesses. Their model included the concepts of "dynamics," responsible for the variation of the

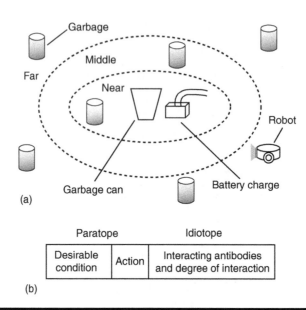

(a)

Paratope		Idiotope
Desirable condition	Action	Interacting antibodies and degree of interaction

(b)

Figure 7.9 **(a) Environment for testing the consensus-making algorithm based on immune networks, (b) definition of antibody.**

concentration level of antibodies, and "metadynamics," which maintained the appropriate repertoire of antibodies. This work was then extended by the authors to attempt in creating a more emergent behavior within the network of robots (Kondro et al., 1998) by the introduction of genetic operators.

KrishnaKumar et al. (1995) and KrishnaKumar and Neidhoefer (1997a, 1997b, 1999) proposed an "immunized computational systems" (ICS). They successfully tested ICS on an autonomous aircraft control problem. Ootsuki and Sekiguchi (1999) suggested a method for determining control sequences of a sequential control plant based on the immune system. They used Petri nets with control sequences equal to the firing sequences of a Petri net.

Attempts have been made to apply the immune network idea to control large populations of robots to have some form of a self-organizing group behavior. Toma et al. (1999) attempted to create a group of robots, which self-organized to search for food without any global control mechanism. The authors used B cells to represent a robot and its stimulation is calculated according to its performance. Each robot carried a record of its degree of success in collecting food, while neighboring robots compared their success and strategies and stimulated and suppressed each other accordingly. This work was advanced by Mitsumoto et al. (1996) with the introduction of the idea of B cell cloning to represent messages to other robots. If a robot is achieving the work, then it receives more stimulation. After a certain stimulation threshold, it produces clone B cells. Also, Mitsumoto et al. (1997) and Lee and Sim (1997) used metaphors based on T cell behavior in robotics.

Wantanabe et al. (1998) developed a work on decentralized adaptive-control mechanism for the walking behavior of a six-legged robot. By using the idea of B cells and immune networks, where a B cell is considered to be a leg in the robot and the immune network a mechanism by which legs communicate with each other, they proposed a system that can learn how to control the walking motion of the robot. Each B cell records the level of antibody concentration. The B cell with the highest concentration is the leg that swings forward.

Singh and Thayer (2001) proposed Immunology-derived Distributed Autonomous Robotics Architecture (IDARA)—a self-optimizing and dynamic robot control architecture. IDARA distributes tasks so that routine actions are refined and followed by more specific responses. They use a multitiered response layer equating it to the response pattern of the innate immune system. Therefore, initial agents can fail and pass on because they fail to successive levels of control, then a more suitable agent is evolved. Basically, the idea is that many agents respond to one problem and the failure of one helps improve the next agent—the idea of innate immunity before the optimum agent can be evolved. Traditional communication in distributed systems has a significantly high cost. They use computer simulation of a "self-healing" mobile minefield having up to 7500 mines and 2750 robots and use a multitiered response layer equating it to the response pattern of the immune system.

Lau and Ko (2007) proposed a robotic search and rescue system based on an immune-control framework, called general suppression control framework (GSCF). GSCF is based on the suppression mechanism of immune cells. A decentralized system based on GSCF to assist a search and rescue a robot system to communicate and navigate in unstructured disaster-affected areas is developed. The robot system consists of two robots and one operator console, but can be extended to a higher number of robots. GSCF is a modular system that consists of five major components: affinity evaluator, cell differentiator, cell reactor, suppression modulator, and the local environment. Thus, autonomous T cells that continuously react to the changing environment and affect other cells in the environment are modeled.

Lee et al. (2007) used clonal selection algorithms in controlling autonomous underwater vehicles. Particularly, clonal selection is used to tune control parameters—K_p, K_D, and K_I—of proportional integral derivative (PID) controllers. The proposed approach was compared with the Ziegler–Nichols (Z–N) technique with respect to the settling time, overshoot and an affinity in submerging underwater, and turning the yaw angle through simulation. The immune approach is more efficient than the Z–N technique in submerging and turning the yaw angle.

7.5 Application in Fault Detection and Diagnosis

The field of fault diagnosis needs to accurately predict or recover from faults occurring in plants, machines such as refrigeration systems, communications such as telephone systems, and transportations such as aircrafts. Bersini and Varela (1990) used learning vector quantization (LVQ) to determine a correlation between two sensors from their outputs when they work properly. Each sensor is equated to a B cell in an immune network, and sensors test one another's outputs to see whether or not they are normal using the extracted correlations. Here, reliability of the sensor is used *in lieu* of the similarity to neighbors.

In the field of diagnosis, there has also been some interest in creating distributed diagnostic systems. Kayama et al. (1995) initially proposed a parallel-distributed diagnostic algorithm. The authors compared their algorithm to that of an immune network due to its distributed operation, and the systems emergent cooperative behavior between sensors. This work was then continued by Ishida (1990, 1996). Active diagnosis continually monitors for consistency between the current states of the system with respect to the normal state. Each sensor can be equated with a B cell, connected through the immune network with each sensor maintaining a time-variant record of sensory reliability, thus creating a dynamic system. This work differs from the aforementioned in the way in which the reliability of each sensor is calculated.

An AIS technique was applied to refrigerated cabinets in supermarkets to detect the early symptoms of icing up. Taylor and Corne (2003) used in-cabinet temperature data to predict faults from the pattern of temperature over time. This technique used r-bits matching rule in conjugation with a specialized differential encoding of data to spot fault patterns in a time-series temperature data from supermarket freezer cabinets.

An aircraft fault-detection system, called multilevel immune learning detection (MILD), was developed (Dasgupta et al., 2004) to detect a broad spectrum of known as well as unforeseen faults. Empirical study was conducted with datasets collected through simulated failure conditions using National Aeronautics and Space Administration (NASA) Ames C-17 flight simulator. Three sets of in-flight sensory information—namely, body-axes roll rate, pitch rate, and yaw rate were considered to detect five different simulated faults: one for engine, two for the tails, and two for the wings. The MILD implemented a real-valued negative selection (RNS) algorithm, where a small number of specialized detectors (as signatures of known failure conditions) and a set of generalized detectors (for unknown or possible faults) are generated. Once the fault is detected and identified, an adaptive control system would use this detection information to stabilize the aircraft by utilizing available resources (control surfaces). Experiments were performed with datasets collected under normal and various simulated failure conditions using a piloted motion-based NASA simulation facility. A snapshot of a running MILD is shown in Figure 7.10.

An artificial immune regulation (AIR) scheme was proposed and integrated into an immune model-based fault detection approach for fault diagnosis (Luh et al., 2004). This system generated residuals that contained information about the faults. However, various disturbances and errors caused residuals to become nonzero, thus interfering with detection of faults. The AIR scheme produced a set of memory B cells whose amount depended on several chemical rate constants.

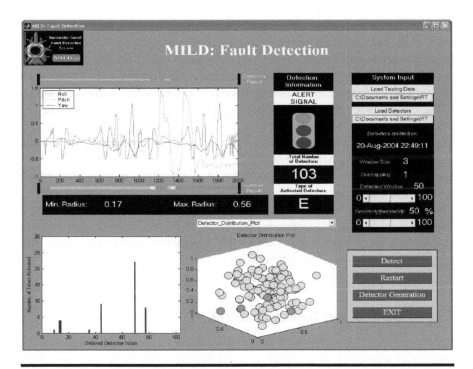

Figure 7.10 **The performance of MILD when tested with "full-tail failure" data, where this type of fault is manifested in pitch error rate (starting at the 1200th time step). The graph also shows the number of detectors activated (lower-bar chart) as significant deviations in data patterns appear. The bar chart shows the arrangement of the detectors with increased radius.**

This helped in classifying the residuals into distinct patterns, denoting different faulty situations.

An algorithm was proposed by Pinto et al. (2005) for detecting faults in telephone systems based on DT in immunology, which is guided by the principle that the presence or absence of secondary signals determines responsiveness or tolerance. Each call in this fault-detection system is represented by an antigen composed of linear attributes: origin, destination, duration of calls, and a nominal attribute. Two signal levels were identified: signal 1 for perceiving the presence of the antigen and signal 2 for costimulation by using the noncompleted call rate. Signal 2 was responsible for alarming a danger situation. Detector death, detector deactivation, detector population renewal, and a voting routine were significantly employed in this work.

Guzella et al. (2007) presented an immune-inspired approach for fault detection called dynamic effector regulatory algorithm (DERA). The proposed approach integrates the role of regulatory T cells in control and signaling between cells. In DERA, new components of the immune system such as cytokines and regulatory cells are incorporated in the model. DERA uses a population of regulatory and effector cells and combines both positive and negative detection; it also keeps track of the concentration of two cytokines in the environment. It is based on the assumption that there must be an interaction between cells in the population before determining whether an antigen belongs to self or nonself. The system possesses a memory that is represented by cytokine concentrations such that the classification of an antigen depends on the responses against recently classified instances. The system does not include clonal selection, thus the memory is not antigen specific. Unlike NS algorithms, which look for a total coverage of the nonself space, DERA searches for an appropriate distribution of effector and regulatory cells throughout the space. By combining both regulatory and effector cells, to recognize normal and abnormal operation, respectively, DERA's dynamic behavior mediated by cytokines is able to indicate the severity of a fault. The proposed approach was tested on the DADAMICS fault-detection benchmark problem, and it was able to attain considerably lower false-positives than other approaches, because regulatory cells suppress the activation of effector cells.

7.6 Application to Scheduling

Creating optimal schedules in a constantly changing environment is not easy. The purpose of scheduling is to allocate a set of limited resources to tasks over time. Ishida (1997) and Mori et al. (1994) proposed and developed an immune algorithm that can create adaptive scheduling system based on the metaphors of somatic hypermutation and immune network theory. Mori et al. (1994) built on this immune algorithm by addressing the issue of batch sizes and combinations of sequence orders, which optimized objective functions. In these works, antigens are considered as input data or disturbances in the optimization problem, and antibodies are considered as possible schedules. Proliferation of the antibodies is controlled by an immune network metaphor where stimulation and suppression are modeled in the algorithm. This assists in the control of antibody (or new solution) production. The T cell effect in this algorithm is ignored. The authors claim that their algorithm is an effective optimization algorithm for scheduling and was shown to be good at finding optimal schedules. The application of this algorithm to a dynamically changing environment has been attempted by Mori et al. (1998). Here, the authors considered antibodies as a single schedule and antigens to be possible changes to the schedule. Their system produced a set of antibodies (schedules) that can cover the whole range of possible changes in the antigen set.

An AIS was utilized by Coello et al. (2003) to solve job-shop scheduling problems (JSSP) using clonal selection, hypermutation, and an antibody library to construct solutions. The purpose of JSSP is to find an optimum schedule that gives the minimum duration to complete all the jobs (n jobs for m machines). It is an optimization problem for particular objectives where certain criteria are met during the assignment. A permutation representation (extensions of CLONALG) is adapted wherein an antibody represents schedule, whereas an antigen holds information on the set of expected arrival dates for each job to the shop. A library of antibodies builds new solutions to the problem.

Ong et al. (2005), in a similar work called "ClonaFLEX," intends to solve the flexible JSSP with recirculation. It employs self-initiated antibody initialization, suitable antibody mutation rates based on their affinities, and a novel distribution of elite pools to produce antibodies. The possible job schedules are modeled as antibodies. The search process is repeatedly carried out and the information gained in each generation is used as feedback to conserve and propagate good features. ClonaFLEX also employs parallel search for optimizing time.

7.7 AIS in Data Mining

Timmis and Knight (2001) wrote a chapter on the concepts of artificial immune systems, particularly on artificial immune networks (AINE), which is a machine-learning algorithm, based on immune network theory as applicable to the field of data mining. The self-organizing nature of B cell network can be used as an efficient clustering tool. The idea of repertoire completeness is achieved by making a certain receptor surface match not only to an exact complementary string, but also to some variations of it, that is, a ball of recognition. The B cell receptors can serve as cluster centers, which will suitably self-organize.

Hunt and Cooke (1996) attempted to apply an immune-based model to data mining by creating a system that could help in the customer-profiling domain. Each B cell object contained customer profile data such as marital status, ownership of cars, and bank account details.

Serapião et al. (2007) used an artificial immune system for the classification of petroleum well drilling operations. Particularly, two approaches based on CLONALG and parallel AIRS2 were developed. They implemented a system, which takes advantage of information collected by mud-logging techniques during well-drilling operations. Mud-logging systems operations collect two types of information: formation samples (shale-shaker samples) and mechanical parameters related to the drilling operation. AIRS2 is a bone marrow clonal selection type of immune algorithm. AIRS2, as in CLONALG, develops a set of memory cells that represents the training data environment. Also, AIRS2 uses affinity maturation and somatic hypermutation. It works on two stages: evolving candidate memory cells and determining whether they should be added to the pool of memory cells or not. Once the training routine is performed, AIRS2 classifies instances using k-nearest neighbor (k-NN) on the set of developed memory cells. Thus, AIRS2 first learns the input space through a clustering process and then uses k-NN on the cluster representatives for classification. The reported results showed that imbalanced real mud-logging data has large impact on the classification performance of the AIS classifiers; they achieve high precision on predominant classes, but lower classification precision on classes with fewer samples.

7.7.1 Applications in Web Mining

In a conceptual paper, Secker et al. (2003) investigated the relevance of DT to Web mining. An adaptive mailbox filter is presented, which essentially employs a dynamical classification task. This system accepts or temporarily ignores incoming e-mails depending on an importance measure decided by the user at a specific instant of time. An antigen represents a processed original e-mail along with its class. Clonal selection and mutation evolve the antibody set, which change and update (including culling) over time reflecting users' changing preferences and the changing nature of received e-mails. The authors state the ultimate idea inspired by the combined artificial tissues capable of releasing artificial danger signals.

Nasraoui et al. (2002) proposed the fuzzy artificial recognition ball (ARB), which represents a fuzzy set over the domain of discourse consisting of the training dataset, as an improvement of the original ARB. The final fuzzy ARB population can be consolidated by a crossover of randomly exchanging chromosomes, or by any other reasonable aggregation such as arithmetically averaging. Synthesized data and Web usage data are mined as the target of this method. For Web usage, the final merged ARBs correspond to typical profiles for the users accessing a given Web site. The average attributes reflect the relevance of the individual URLs to the combined ARBs.

Nasraoui et al. (2006) proposed a scalable immune-inspired clustering methodology to continuously learn and adapt to new incoming patterns in Web mining. In this work, the Web server plays the role of the human body, and the incoming requests play the role of foreign antigens/bacteria/viruses that need to be detected by the proposed immune-based clustering technique. Hence, this immune algorithm is used to continuously perform clustering of the incoming noisy data. The authors claim that the proposed approach exhibits superior learning abilities while requiring modest memory and computational costs. An important advantage of this method is its adaptation to the dynamic environment that characterizes several applications, particularly in mining data streams. The performance of the proposed approach is tested on mining user profiles from Web clickstream data in a single pass under different usage trend-sequencing scenarios.

7.7.2 Application in Anomaly Detection

Dasgupta and Forrest (1996) and Dasgupta and Gonzalez (2002) propose the use of the NS algorithm to the application of detecting anomalies in general time series data. A number of experiments were performed using Mackey–Glass time series and other datasets (algorithmic steps in Figure 7.11).

In most of the works on anomaly detection, a sliding window scheme was used for data preprocessing, which is illustrated in Figure 7.12.

Gonzalez et al. (2002) implemented an RNS and compared against an unsupervised learning algorithm using different datasets in anomaly detection. A further

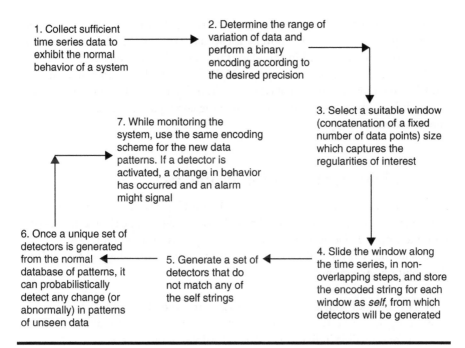

Figure 7.11   Different steps in implementing anomaly/novelty detection.

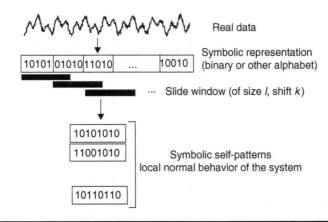

Figure 7.12   Preprocessing of time series date in anomaly detection.

improved version called randomized RNS (RRNS; Gonzalez et al., 2003) produced a good estimate of the optimal number of detectors with maximization of the non-self coverage done through an optimization algorithm with proved convergence properties. This was based on a type of randomized algorithms called "Monte Carlo methods." Specifically, it uses Monte Carlo integration and simulated annealing.

A framework called multilevel immune learning algorithm (MILA) is proposed by Dasgupta et al. (2003). The novelty is twofold: (1) More mechanisms of natural immune system are utilized in this algorithm, including T helper cells, T suppressor, B cells, and antigen-presenting cells (APCs). This makes it different from other models using only the concept of either B cells or T cells. (2) This algorithm detects in a multilevel, multiresolution fashion, making large space to explore efficiently for anomaly-detection applications.

7.8   Solving Optimization Problems

Hajela and Lee (1996), Hajela et al. (1997), and Hajela and Yoo (1999) described the implementation of a constrained genetic search to simulate the mechanics of an immune-inspired algorithm to solve engineering-optimization problems.

Coello and Cortes (2002) used NS algorithm to handle the problem of infeasible solution. Accordingly, they assumed

■ Feasible individuals = Ag and infeasible individuals = Ab
■ Antigen–antibody interaction (clone mutate and increase affinity) used to make infeasible individuals move to feasible solution space

Endoh et al. (1998) and Toma et al. (1999) proposed an adaptive optimization algorithm for the traveling salesman problem. This approach is based on the immune network model and major histocompatibility complex (MHC) peptide presentation. Here, the immune network principles were used to simulate adaptive behavior of agents, various concepts such as MHC to induce competitive behavior among agents, T cells as control behavior, and B cells as produce behavior (Table 7.1 shows the specific mappings and the schematics diagram [Figure 7.13] illustrates the corresponding components).

Table 7.1   Immune Cells and Molecules and Their Roles in the n-TSP Problem Solving

Immune System	Role in the n-TSP problem
Antigen	Contains information about the cities and salesmen
Macrophage	Selects the city number that the salesman agent must visit
T cells	Help the activation of B cell
B cells	Produce antibodies
Antibody	Performs the behavior of an agent

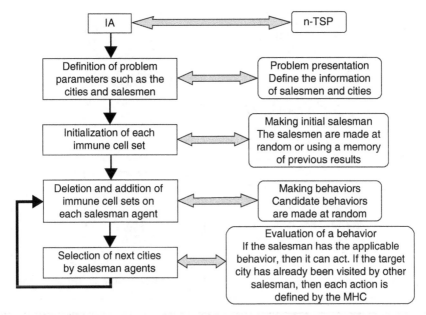

Proposed immune algorithm (IA) for solving the n-travelling salesman problem (TSP). Each immune cell set is composed of three kinds of cells, called a macrophage, a B, and a T cell.

Figure 7.13   Corresponding components of immune algorithm and n-TSP.

7.9   Other Applications

7.9.1   Developing Associative Memories

Researchers (Smith et al., 1996) argued that the immunological memory is a member of the family of sparsely distributed memories, and it derives associative and robust properties from a sparse and distributed nature of sampling.

Figure 7.14 illustrates the formation of immune memory (as the concentration level of various immune cells) during the primary and secondary responses.

AINs have been applied to create an associative memory model (Singh and Thayer, 2001). Associative memory is used to remember patterns and enable fast and effective recall of those patterns. The authors implemented two mechanisms defined by Abbattista et al. (1996), namely, the immune system metadynamics and the immune recruitment mechanism. A population of points in the space is defined, which compete to recruit items from the training population. These result in clustering areas on the surface space that, in effect, store the patterns being learned.

7.9.2   Applications in Games

Many AIS researchers (Varela et al., 1988) talked about the inherent capability of the immune networks for machine learning. Perelson (1989) and Cooke and Hunt (1995)

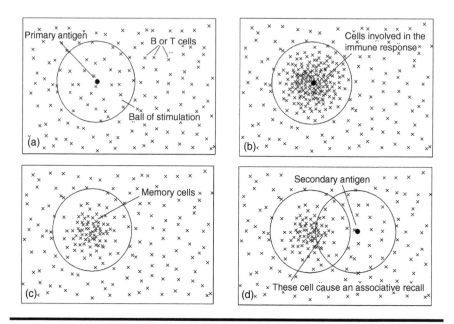

Figure 7.14 **Modeling immune memory as associative memory. (From Smith D. J., S. Forrest and A. S. Perelson,** *Artificial Immune Systems and Their Applications, 1999, The International Conference on Multi-Agent Systems,* **Workshop Notes, Kyoto, 1996, 62–70.)**

made some of the earliest attempts to use immune system metaphors in machine learning. They applied AINE (described in Chapter 5) to the problem of playing knots and crosses. In this system, each B cell corresponded to a particular board state containing a nine-digit antibody. The good moves from one state to another meant that those two B cells would have strong affinity or a connection in the B cell network. Later, this group also applied this algorithm to the domain of case-based reasoning. In this system, each case is represented by a B cell object and the case memory is built with the B cell network, with similar cases being linked together. The memory was self-organizing in nature.

7.9.3 Applications in Software Testing

May et al. (2007) presented an immune-inspired system based on CLONALG for software test data. Generated test datasets are evaluated using the mutation testing adequacy criteria and are used to direct the search of new tests. Mutation testing generates versions of a program containing simple faults and then finds tests to indicate the program's symptoms. The developed immune system for mutation testing is based on the clonal selection algorithm. A modified version of CLONALG to adapt it to the mutation testing problem, removing the concept of a memory individual per antigen, and allowing many memory individuals to contribute to antigen recognition are developed. This immune approach is compared with a genetic-based approach, and it showed to be consistently more efficient, generating higher mutation scoring test sets at a less computational cost.

7.10 Hybrid Approaches

7.10.1 *Application in Neural Networks*

In Dasgupta and Forrest (1999a, 1999b), ongoing work using simulated annealing and immune metaphors applied to the problem of finding good initialization vector for neural networks is presented. The strategy needs no prior knowledge about the problem except the assumption that the error surface has multiple local optima. But a good region to search needs to be sampled for a good solution. In simulated annealing for diversity (SAND), an antibody is a possible solution for the weight vector of a given neuron in a single layer of the network. Antigens (the training data) are ignored here. The goal is to maximize the Ab–Ab distance so that similarity in the population is reduced. Affinity is measured as the Euclidean proximity between two points in an n-dimensional shape space. The energy measure to be optimized is the sum of the Euclidean distances among all vectors that represent the Ab population. They stop the search process whenever the distribution of the Ab population reaches a close-to-uniform distribution.

7.10.2 *Applications in Genetic Algorithms*

To address the issue of designing a GA with improved convergence characteristics, particularly in the field of design constraints, Slavov and Nikolaev (1998) proposed a GA simulation of the immune system. GAs have been found to be very sensitive to the choice of algorithm parameters when applied to design constraints. The authors used the idea of antibody–antigen binding. The fitness of a solution is not only dependent on the objective function value and the design constraints, as it would be in a traditional GA, but also on how well the solution matches the best solution. The algorithm then selects these better solutions to adapt, thus leading to a higher convergence rate when compared to a traditional GA. Hajela et al. (1997) adopt a more generic approach to the adaptive problem solving by the use of the immune network metaphor such as B cells, T cells, macrophages, and the MHC. The immune algorithm is used to produce adaptive behaviors of agents. Hajela et al. also experimented by removing the interaction of the T cell in the searching algorithm, and present convincing results that the effect of the T cell on performance is significant as the solutions found with using the T cell result in lower-cost solutions overall.

Other similar applications of the immune network metaphor for multimodal function optimization can be found in Toma et al. (2000), Fukuda et al. (1999), and Mori et al. (1998). Here, the authors use somatic hypermutation and immune network theory to create and sustain a diverse set of possible solutions in the search space and combine it with traditional GAs.

Endoh et al. (1998) used immune system ideas to handle constraints in GAs. Traditionally, the way to implement the constraints true to real life has been the use of penalty functions to the fitness evaluation process so that the solutions are guided away from infeasible regions of the solution landscape. However, good penalty factors are difficult to define. This technique is based on the NS algorithm. It generates a random population and denotes the feasible individuals as the antigens and the infeasible individuals as the antibodies. The idea is that antibodies learn from antigens to be closer to them or the infeasible individuals are motivated to become feasible by their exposure to the feasible part of the population. The evolution is carried out using standard GA operators.

Mori et al. (1998) looks at applying immune system metaphors to extended GAs for search optimization problems. Here, the authors propose an extension to a standard GA by the inclusion of immune system metaphors of B cells by using a combination of memory and suppression cells. This variation of the algorithm creates a memory of the best cases for searching, allowing the reinforcement of good solutions within the search space, and to use those good solutions for further exploration. This work was then extended by Coello and Cortes (2002).

7.11　Summary

IC emerged in the 1990s as a new paradigm in artificial intelligence (AI), and has earned its position on the map of soft computing. This chapter summarizes the applications of artificial immune systems in various science and engineering domains. A survey of some of the applications in this emerging field of artificial immune systems has been reported, which include computer security, anomaly detection, pattern recognition, data mining, adaptive control, fault detection, and many others.

To apply an immune model to solve a particular problem from a specific domain, one should select the immune algorithm according to the type of problem that is being solved. Then, identify the elements involved in the problem and how they can be modeled as entities in the particular immune model. To model such entities, a representation for each of these elements should be chosen, specifically, a string representation: integer, real-valued vector representation, or a hybrid representation. Subsequently, appropriate affinity (distance) measure to determine corresponding matching rules should be defined. Then the immune algorithm that will be used to generate a set of suitable entities providing a good solution to the problem at hand should be selected.

The following issues concerning the type of the problems and the property of the training data are important in any analysis of NS algorithms:

■ *Frequency-reflecting data.* The distribution probability of data is crucial to evaluate the success of the learning algorithm.

- *Noisy data*. Is there a need of mechanisms to deal with noise and outliners, or any other "bad" self-samples? In real applications, it seems necessary in most scenarios.
- Recognizing each abnormal data or reacting only when the same anomaly happens multiple times. The latter is more reasonable in applications such as network intrusion detection or fault detection.
- *Completeness of the self-samples*. If all self-patterns are represented in the training data, we call it "complete." It is usually not the case for real-valued representation, but it is true for many applications using string representation.
- *Dynamic data*. This refers to the training data that are not stationary (self-set changes over time).
- *Distributed data*. In this case, the self-set is distributed or too large to observe completely.

Forrest et al. (1994) mentioned that the method relies on the fact that the data is not corrupted when the detectors are generated. This reflects the idea that the self-samples are at least considered correct regardless of whether they are complete or not. Even if the self-samples are complete as well as correct, NS algorithms are still probabilistic in most methods, implying that they may not achieve perfect coverage. The goal is to have a small number of detectors that are capable of detecting a relatively large portion of nonself space. In brief, the algorithms should depend on the data properties as little as possible, but the common assumptions are important for a plausible comparison.

7.12   Review Questions

1. Order and explain the steps necessary to apply an immunological model to solve a particular problem.
   a. Identify the elements involved in the problem
   b. Identify how elements can be modeled as entities
   c. Choose a representation for each one of the elements involved in the problem
   d. Define appropriate affinity measures
   e. Decide what immune algorithms will be used
2. Develop an immunity-based multiagent system to play robot soccer. Describe the original solution based on the steps shown in Figure 7.1.
3. Explain the advantages of using AINEs over other techniques in each of the following applications:
   a. Machine learning
   b. GAs
   c. Robot behavior and control
   d. Fault diagnosis

   e. Scheduling
   f. Computer security
   g. Anomaly detection
   h. Neural networks
   i. Data mining
4. Study other research areas where AINEs have been successfully used. Identify the main features of these approaches.

References

Abbattista, F., G. Di Santo, G. Di Gioia and M. Fanelli. An associative memory based on the immune networks. *Proceedings of the International Conference on Neural Networks*, Washington, DC, 1996.

Aickelin, U. and J. Greensmith. Sensing danger: Innate immunology for intrusion detection. *Inform. Secur. Tech. Rep.*, 12(4), 218–227, 2007.

Aicklen, U., P. Bentley, S. Cayzer, J. Kim and J. McLeod. Danger theory: The link between ais and ids. *The Proceedings of the Third International Conference, ICARIS 2003 on Artificial Immune Systems*, Edinburgh, pp. 156–167, 2003.

Balthrop, J., S. Forrest and M. Glickman. Revisiting lisys: Parameters and normal behavior. *Proceedings of the Congress on Evolutionary Computation*, Honolulu, HI, pp. 1045–1050, 2002.

Bersini, H. and F. Varela. Hints for adaptive problem solving gleaned from immune network. In H. P. Schwefel and H. M'hlenbein (Eds.), *Parallel Problem Solving from Nature*, Springer, Berlin, pp. 343–354, 1990.

Carter, J. H. The immune system as a model for pattern recognition and classification. *J. Am. Med. Inf. Assoc.*, 7(1), 28–41, 2000.

Coello, C. A. C. and N. C. Cortes. Use of emulations of the immune system to handle constraints in evolutionary algorithms. *Proceedings of ANNIE 2002*, St. Louis, MO, 2002.

Coello, C. A. C., D. C. Rivera and N. C. Cortes. Use of an artificial immune system for job shop scheduling. *The Proceedings of the Second International Conference, ICARIS 2003 on Artificial Immune Systems*, Edinburgh, September 2003.

Cooke, D. E. and J. E. Hunt. Recognizing promoter sequences using an artificial immune system. *Proceedings Intelligent Systems in Molecular Biology (ISMB'95)*, AAAI Press, Cambridge, U.K., pp. 89–97, 1995.

Dasgupta, D. Using immunological principles in anomaly detection. *Proceedings of the Artificial Neural Networks in Engineering (ANNIE'96)*, St. Louis, MO, November 10–13, 1996.

Dasgupta, D. Immunity-based intrusion detection systems: A general framework. *The Proceedings of the 22nd National Information Systems Security Conference (NISSC)*, Arlington, VA, October 18–21, 1999.

Dasgupta, D. and S. Forrest. Novelty detection in time series data using ideas from immunology. *ISCA 5th International Conference on Intelligent Systems*, Reno, Nevada, June 19–21, 1996.

Dasgupta, D. and S. Forrest. An anomaly detection algorithm inspired by the immune system. *Artificial Immune Systems and Their Applications*, Springer, Berlin, pp. 262–277, 1999a.

Dasgupta, D. and S. Forrest. *New Ideas in Optimization*, McGraw-Hill, New York, 1999b.

Dasgupta, D. and F. Gonzalez. An immunity-based technique to characterize intrusion in computer networks. *IEEE Trans. Evol. Compt.*, 6(3), 1081–1088, 2002.

Dasgupta, D., K. KrishnaKumar, D. Wong and M. Berry. Negative selection algorithm for aircraft fault detection. *The Proceedings of the Third International Conference, ICARIS 2004 on Artificial Immune Systems*, Catania, Sicily, September 2004.

Dasgupta, D., S. Yu and N. S. Majumdar. MILA—multilevel immune learning algorithm. *Proceedings of the Genetic and Evolutionary Computation Conference (GECCO 2003)*, LNCS 2723, Springer, Chicago, IL, July 12–16, pp. 183–194, 2003.

D'haeseleer, P., S. Forrest and P. Helman. An immunological approach to change detection: Algorithms, analysis, and implications. *Proceedings of the 1996 IEEE Symposium on Computer Security and Privacy*, IEEE Computer Society Press, Los Alamitos, CA, pp. 110–119, 1996.

Endoh, S., N. Tom and K. Yamada. Immune algorithm for n-tsp. *Proceedings of IEEE International Conference on Systems and Man and Cybernetics (SMC)*. IEEE, San Diego, CA, pp. 3844–3849, 1998.

Forrest, S., S. Hofmeyr and A. Somayaji. Computer immunology. *Commn. ACM*, 40(10), 88–96, 1997.

Forrest, S., S. A. Hofmeyr, A. Somayaji and T. A. Longstaff. A sense of self for Unix processes. *Proceedings of 1996 IEEE Symposium on Computer Security and Privacy*, Oakland, CA, 1996.

Forrest, S., A. S. Perelson, L. Allen, and R. Cherukuri. Self-nonself discrimination in a computer. *Proceedings of the 1994 IEEE Symposium on Research in Security and Privacy*, IEEE Computer Society Press, Los Alamitos, CA, 1994.

Fukuda, T., K. Mori and M. Tsukiyama. Parallel search for multimodal function optimization with diversity and learning of immune algorithm, In D. Dasgupta (Ed.), *Artificial Immune Systems and Their Applications*, Springer, Berlin, pp. 210–219, 1999.

González, F. An immunity-based technique to characterize intrusions in computer networks. *IEEE Trans. Evol. Compt.*, 6(3), 281–291, 2002.

González, F., D. Dasgupta and R. Kozma. Combining negative selection and classification techniques for anomaly detection. *Proceedings of the 2002 Congress on Evolutionary Computation CEC2002*, D. B. Fogel, M. A. El-Sharkawi, X. Yao, G. Greenwood, H. Iba, P. Marrow and M. Shackleton, (Eds.), IEEE Press, Honolulu, HI, pp. 705–710, May 2002.

Gonzalez, F., D. Dasgupta and L. F. Nino. A randomized real-valued negative selection algorithm. *The Proceedings of the Second International Conference, ICARIS 2003 on Artificial Immune Systems*, Edinburgh, September 2003.

Greensmith, J., U. Aickelin and J. Twycross. Articulation and clarification of the dendritic cell algorithm. *ICARIS-06*, LNCS 4163, Oeiras, Portugal, pp. 404–417, 2006.

Guzella, T. S., T. A. Mota-Santos and W. M. Caminhas. A novel immune inspired approach to fault detection. *Lecture Notes in Computer Science, Proceedings of ICARIS 2007*, Springer, Santos, 2007.

Hajela, P. and J. Lee. Constrained genetic search via schema adaptation. In D. Come, M. Dorigo, F. Glover, D. Dasgupta, P. Mascato, R. Poli and K. V. Price (Eds.,), An immune network Solution. *Struct. Optimization*, 12(1), 11–15, 1996.

Hajela, P. and J. S. Yoo. Immune network modeling in design optimization. In D. Corne, M. Dorigo, F. Glover, D. Dasgupta, P. Moscato, R. Poli and K. V. Price (Eds.), *New Ideas in Optimization*, McGraw-Hill, New York, pp. 203–215, 1999.

Hajela, P., J. Yoo and J. Lee. GA based simulation of immune networks—applications in structural optimization. *J. Eng. Optimization*, 29, 131–149, 1997.

Hofmeyr, S. A. and S. Forrest. Immunity by Design: An Artificial Immune System. *Proceedings of GECCO Conference*, San Francisco, CA, 1999.

Hofmeyr, S. A. and S. Forrest. Architecture for an artificial immune system, *Envol. Comput.* 7(1), 45–68, 2000.

Hofmeyr, S. A., A. Somayaji and S. Forrest. Intrusion detection using sequences of system calls. *J. Comput. Secur.*, 6, pp. 151–180, 1998.

Hunt, J. E. and D. E. Cooke. An adaptive and distributed learning system based on the immune system. *Proceedings of the IEEE International Conference on Systems Man and Cybernetics*, Vancouver, BC, pp. 2494–2499, 1995.

Hunt, J. E. and D. E. Cooke. Learning using an artificial immune system. *J. Network Comput. Appl.: Spec. Issue Intell. Syst.: Des. Appl.*, 19, 189–212, 1996.

Hunt, J. E. and A. Fellows. Introducing an immune response into a CBR system for data mining. *BCS ESG'96 Conference and Published as Research and Development in Expert Systems XIII*, 1996.

Hunt, J., J. Timmis, D. Cooke, M. Neal and C. King. Jisys: The development of an artificial immune system for real world applications. In D. Dasgupta (Ed.), *Applications of Artificial Immune Systems*, Springer, Berlin, pp. 157–186, 1999.

Ishida, Y. Fully distributed diagnosis by PDP learning algorithm: Towards immune network pdp model. *Proceedings of International Joint Conference on Neural Networks*, San Diego, CA, pp. 777–782, 1990.

Ishida, Y. Distributed and autonomous sensing based on immune network. *Proceedings of Artificial Life and Robotics*, Beppu, Japan, AAAI Press, pp. 214–217, 1996.

Ishida, Y. Active diagnosis by self-organization: An approach by the immune network metaphor. *Proceedings of the International Joint Conference on Artificial Intelligence.* Nagoya, Japan, IEEE, pp. 1084–1089, 1997.

Ishiguro, A., S. Ichikawa, T. Shibat and Y. Uchikawa. Modernationsim in the immune system: Gait acquisition of a legged robot using the metadymics function. *Proceedings of IEEE International Conference on Systems and Man and Cybernetics (SMC)*, IEEE, San Diego, CA, pp. 3827–3832, 1998.

Ishiguro, A., T. Kondo, Y. Watanabe, Y. Shirai and Y. Uchikawa. Immunoid: A robot with a decentralized consensus-making mechanism based on the immune system. *ICMAS Workshop on Immunity-Based Systems*, Kyoto, Japan, pp. 82–92, December 1996.

Kayama, M., Y. Sugita, Y. Morooka and S. Fukuodka. Distributed diagnosis system combining the immune network and learning vector quantization. *Proceedings of IEEE 21st International Conference on Industrial Electronics and Control and Instrumentation*, IEEE, Orlando, FL, pp. 1531–1536, 1995.

Kephart, J. O. A biologically inspired immune system for computers. In R. A. Brooks and P. Maes, (Eds.), *Artificial Life IV. Proceedings of the 4th International Workshop on the Synthesis and Simulation of Living Systems*, MIT Press, Cambridge, MA, pp. 130–139, 1994.

Kim, J. and P. J. Bentley. An evaluation of negative selection in an artificial immune system for network intrusion detection. *Proceedings of the Genetice and Evolutionary Computation Conference (GECCO 2001)*, San Francisco, CA, 2001.

Kim, J., W. O. Wilson, U. Aickelin and J. McLeod. Cooperative automated worm response and detection immune algorithm (CARDINAL) inspired by T-cell immunity and tolerance. *The Proceedings of the Fourth International Conference on Artificial Immune Systems*, Banff, Alberta, Canada, pp. 168–181, August 2005.

Kondro, T., A. Ishiguro, Y. Wantanabe and Y. Uchikawa. Evolutionary construction of an immune network based behavior arbitration mechanism for autonomous mobile robots. *Electr. Eng. Jpn.*, 123(3), 1–10, 1998.

KrishnaKumar, K. and J. C. Neidhoefer. Immunized Neurocontrol. *Expert Syst. Appl.*, 13(3), 201–214, 1997a.

KrishnaKumar, K. and J. C. Neidhoefer. *Immunized Adaptive Critics*. ICNN, Houston, TX, June 1997b.

KrishnaKumar, K. and J. C. Neidhoefer. Immunized adaptive critic for an autonomous aircraft control application. In D. Dasgupta (Ed.), *Artificial Immune Systems and Their Applications*, Springer, Berlin, pp. 221–240, 1999.

KrishnaKumar, K., A. Satyadas and J. C. Neidhoefer. An immune system framework for integrating computational intelligence paradigms. *Computational Intelligence, A Dynamic Perspective*, IEEE Press, Washington, 1995.

Lamont, G. B., R. E. Marmelstein and D. A. Van Veldhvizen. A distributed architecture for a self-adaptive computer virus immune system. In D. Corne, M. Dorigo, F. Glover, D. Dasgupta, P. Moscato, R. Poli and K. V. Price (Eds.), *New ideas in optimization*, (edited volume), McGraw-Hill, New York, pp. 167–184, 1999.

Lau, H. Y. K. and A. Ko. An immuno robotic system for humanitarian search and rescue. *Lecture Notes in Computer Science, Proceedings of ICARIS 2007*, Springer, Santos, 2007.

Lee, D., H. Jun and K. Sim. Artificial immune system for realization of co-operative strategies and group behavior in collective autonomous mobile robots. *Proceedings of Fourth International Symposium on Artificial Life and Robotics*, AAAI, Oita, Japan, pp. 232–235, 1999.

Lee, D. and K. Sim. Artificial immune network based cooperative control in collective autonomous mobile robots. *Proceedings of IEEE International Workshop on robot and Human Communication*. Sendai, Japan, IEEE, pp. 58–63, 1997.

Lee, J., M. Roh, J. Lee and D. Lee. Clonal selection algorithms for 6-DOF PID control of autonomous underwater vehicles. *Lecture Notes in Computer Science, Proceedings of ICARIS 2007*, Springer, Santos, 2007.

Luh, G.-C., C.-Y. Wu, W.-C. Cheng. Artificial Immune Regulation (AIR) for model-based fault diagnosis. *The Proceedings of the Third International Conference, ICARIS 2004 on Artificial Immune Systems*, Catania, Sicily, September 2004.

May, P., J. Timmis and K. Mander. Immune and evolutionary approaches to software mutation testing. *Lecture Notes in Computer Science, Proceedings of ICARIS 2007*, Springer, Santos, 2007.

Mitsumoto, N., T. Fukuda, F. Arai and H. Ishihara. Control of distributed autonomous robotic system based on the biologically inspired immunological architecture. *Proceedings of IEEE International Conference on Robotics and Automation*, IEEE, Albuquerque, NM, pp. 3551–3556, 1997.

Mitsumoto, N., T. Fukuda and T. Idogaki. Self-organizing multiple robotic system. *Proceedings of IEEE International Conference on Robotics and Automation*, IEEE, Minneapolis, MN, pp. 1614–1619, 1996.

Mori, K., M. Tsukiyama and T. Fukuda. Immune algorithm and its application to factory load dispatching planning. *JAPAN-U.S.A. Symposium on Flexible Automation*, Ann Arbor, MI, pp. 1343–1346, 1994.

Mori, K., M. Tsukiyama and T. Fukuda. Adaptive scheduling system inspired by immune system. *IEEE International Conference on Systems, Man, and Cybernetics*, San Diego, CA, 1998.

Nasraoui, O., C. Cardona, C. Rojas and F. Gonzalez. Mining evolving user profiles in noisyweb clickstream data with a scalable immune system clustering algorithm. *Comput. Networks*, 50(10), 1488–1512, 2006.

Nasraoui, O., F. Gonzalez and D. Dasgupta. The fuzzy ais: Motivations, basic concepts, and applications to clustering and web profiling. *IEEE International Conference on Fuzzy Systems*, Hawaii, May 12–17, 2002, pp. 711–717.

Neal, M., J. Hunt and J. Timmis. Augmenting an artificial immune network. *Proceedings of International Conference Systems and Man and Cybernetics*, San Diego, CA, IEEE, pp. 3821–3826, 1998.

Ong, Z. X., J. C. Tay and C. K. Kwoh. Applying clonal selection principle to find flexible job-shop schedules. *The Proceedings of the Fourth International Conference, ICARIS 2005 on Artificial Immune Systems*, Banff, Alberta, August 2005.

Ootsuki, T. and T. Sekiguchi. Application of the immune system network concept to sequentialcontrol. *Proceedings of IEEE International Conference on Systems, Man, and Cybernetics*, Tokyo, Japan, Vol. 3, pp. 869–874, 1999.

Perelson, A. S. Immune network theory. *Immunol. Rev.*, 110(10), 5–36, 1989.

Pinto, J. C. L. and Fernando J. Von Zuben. Fault detection algorithm for telephone systems based on the danger theory. *The Proceedings of the Fourth International Conference, ICARIS 2005 on Artificial Immune Systems*, Banff, Alberta, August 2005.

Secker, A., A. A. Freitas and J. Timmis. A danger theory approach to web mining. *The Proceedings of the Second International Conference, ICARIS 2003 on Artificial Immune Systems*, Edinburgh, September 2003.

Serapião, A. B. S., J. Ricardo, P. Mendes and K. Miura. Artificial immune systems for classification of petroleum well drilling operations. *The Proceedings of the 6th International Conference on Artificial Immune Systems (ICARIS)*, Brazil, August 26–29, 2007.

Singh, S. P. N. and S. M. Thayer. Immunology directed methods for distributed robotics: A novel, immunity-based architecture for robust control & coordination. *The Proceedings of SPIE: Mobile Robots XVI*, Vol. 4573, November 2001.

Slavov, V. and N. Nikolaev. Immune network dynamics for inductive problem solving. In A. E. Eiben, T. Back, M. Schoenauer, and H. P. Schwefel (Eds.), *Parallel Problem Solving from Nature*, PPSN V, LNCS-1498, Springer, Berlin, pp. 712–721, 1998.

Smith, D. J., S. Forrest and A. S. Perelson. Immunological memory is associative. *Artificial Immune Systems and Their Applications, 1999, The International Conference on Multi-Agent Systems*, Workshop Notes, Kyoto, Japan, pp. 62–70, 1996.

Somayaji, A., S. Hofmeyr and S. Forrest. Principles of a computer immune system. *1997 New Security Paradigms Workshop*, pp. 75–82, 1998.

Taylor, D. W. and D. W. Corne. An investigation of the negative selection algorithm for fault detection in refrigeration systems. *The Proceedings of the Second International Conference, ICARIS 2003 on Artificial Immune Systems*, Edinburgh, September 2003.

Timmis, J. and T. Knight. Artificial immune systems: Using the immune system as inspiration for Data Mining. In H. A. Abbass, R. A. Sarker and C. S. Newton (Eds.), *Data Mining: A Heuristic Approach*, chapter XI, Group Idea Publishing, pp. 209–230, 2001.

Toma, N., S. Endo and K. Yamada. Immune algorithm with immune network and mhc for adaptive problem solving. *Proceedigs of IEEE International Conference on Systems and Man and Cybernetics (SMC)*, Tokyo, IEEE, 1999.

Toma, N., S. Endo and K. Yamada. The proposal and evaluation of an adaptive memorizing immune algorithm with two memory mechanisms. *J. Jpn. Soc. Artif. Intell.*, 15(6), 1097–1106, 2000.

Twycross, J. Integrated innate and adaptive artificial immune systems applied to process anomaly detection. PhD thesis, School of Computer Science, The University of Nottingham, 2007.

Varela, F., A. Coutinho, B. Dupire and N. Vaz. Cognitive networks: Immune and neural and otherwise. *Theor. Immunol.: Part Two, SFI Stud. Sci. Complexity*, 2, 359–371. 1988.

Wantanabe, Y., A. Ishiguro, Y. Shirai and Y. Uchikawa. Emergent construction of a behavior arbitration mechanism based on the immune system. *Adv. Robotics*, 12(3), 227–242, 1998.

Wantanabe, Y., A. Ishiguro and Y. Uchikawa. Decentralized behavior arbitration mechanism for autonomous mobile robot using immune network. *Artificial Immune Systems and Their Applications*, Springer, Berlin, pp. 187–207, 1999.

Warrender, C., S. Forrest and B. Pearlmutter. Detecting intrusions using system calls: Alternative data models. *1999 IEEE Symposium on security and Privacy*, 1999.

Yeom, K.-W. Immune-inspired algorithm for anomaly detection. *Stud. Comput. Intell. (SCI)*, 57, 129–154, 2007.

Appendix: Indexed Bibliography

Artificial Immune Systems

The field of Artificial Immune Systems (AIS) is becoming more popular and AIS-based works spanning from theoretical modeling and simulation to a wide variety of applications. In particular, some of the references are of synthetic approaches to understand and simulate the biological immune system, and others that develop computational methodologies inspired by the immune system to solve real-world problems. The AIS research group at the University of Memphis headed by Professor Dipankar Dasgupta has been publishing the updated AIS bibliography since 1997. Although this bibliography has been compiled with the utmost care and we tried to make it a complete review of the references in the field, there may be errors in the references we cited and we may have left out some important citations. In either case, we will appreciate any help you give us to update the future versions. All comments, suggestions, and additions are welcome to improve this bibliography. Please send your contributions to Professor Dipankar Dasgupta (dasgupta@memphis.edu). The compilers are also grateful to the researchers who helped us in our literature collection by sending either copies of citations or copies of documents. The authors take no responsibility, however, for any errors, missing information, the contents and quality of the references, nor for the usefulness and the consequences of applying the models or methodologies.

Books/Edited Volumes

1. *Recent Developments in Biologically Inspired Computing*, L. N de Castro and F. J. Von Zuben, (Eds.) Idea Group Incorporation. 2004. ISBN: 1-59140-312-X.
2. *Immunity-Based-Systems: A Design Perspective*, Yoshiteru Ishida, Verlag/Jahr: Springer, 192 p. Berlin 2004. ISBN: 3-540-00896-9.

Last compiled: December 2007 (Regularly updated, and available at: http://ais.cs.memphis.edu/files/papers/

3. *Artificial immune Systems* (Special Issue on of the Journal on Genetic Programming and Evolvable Machines), J. Timmis and P. Bentley (Guest Eds.) Volume 4, No. 4, December 2003.

4. *Perspectives on Adaptation in Natural and Artificial Systems.* L. Booker, S. Forrest, M. Mitchell, and R. Riolo (Eds.), Oxford University Press.

5. *Immunocomputing: Principles and Applications*, A.O. Tarakanov, V.A. Skormin and S.P. Sokolova, Springer-Verlag, 2003. ISBN: 0-387-95533-X.

6. *Artificial Immune Systems* (Special issue of the journal IEEE Transaction on Evolutionary Computation). D. Dasgupta (Guest ed.), Vol. 6, No. 3, June 2002.

7. *Artificial Immune Systems: A New Computational Intelligence Approach,* L. N. de Castro and J. Timmis, Springer-Verlag, Heidelberg, Germany, August 2002. ISBN: 1-85233-594-7.

8. *Sztuczne systemy immunologiczne. Teoria i zastosowania* (Book in Polish). (Artificial Immune Systems. Theory and Applications). S. T. Wierzchon. Akademicka Oficyna Wydawnicza EXIT, Warszawa 2001. ISBN 83-87674-30-3.

9. *Design Principles for Immune System and Other Distributed Autonomous Systems,* Segel and Cohen (Eds). Oxford University Press, 2000.

10. *Artificial Immune Systems and Their Applications*, D. Dasgupta (Ed.) Springer-Verlag. 1999.

Book Sections

1. *Introductory Tutorials in Optimisation, Search and Decision Support Methodology,* E. Burke and G. Kendall (Eds.), Kluwer, 2005.

2. *Towards a danger theory inspired artificial immune system for web mining.* Andrew Secker, Alex Freitas, and Jon Timmis. In A Scime, editor, Web Mining: applications and techniques, pages 145–168. Idea Group, January 2005.

3. *Intelligent Information Systems.* Series: Advances in Soft Computing. Zadeh, Kacprzyk (Eds.) Springer, Verlag, 2000.

4. *New Ideas in Optimization*, D. Corne, M. Dorigo and F. Glover (Eds.), McGraw-Hill, 1999.

PhD Dissertations

1. Dr. Zhou Ji. Dissertation title: *Negative Selection Algorithms: from the Thymus to V-detector.* Department of Computer Science. The University of Memphis, Summer 2006.

2. Albert W. Y. Ko. Dissertation Title: *The Design of an Immunity-based Search and Rescue System for Humanitarian Logistics.* The University of Hong Kong, 2006.

3. Andrew Secker. Dissertation Title: *Artificial Immune Systems for Web Content Mining: Focusing on the Discovery of Interesting Information,* University of Kent, 2006.

4. Thomas Stibor. Dissertation Title: On the Appropriateness of Negative Selection for Anomaly Detection and Network Intrusion Detection. Darmstadt University of Technology, 2006.

5. F. Esponda. Dissertation Title: *Negative Representations of Information*, Ph.D. thesis, University of New Mexico, 2005.

6. Modupe Ayara. Dissertation Title: *An Immune Inspired Approach For Adaptable Error Detection in Embedded Systems*, University of Kent, Canterbury, U.K., 2005.

7. Andrew B. Watkins. Dissertation Title: *Exploiting Immunological Metaphors in the Development of Serial, Parallel, and Distributed Learning Algorithms.* University of Kent, Canterbury, U.K., March 2005.

8. Luis J. Gonzalez. Dissertation Title: *A Self-Adaptive Evolutionary Negative Selection Approach for Anomaly Detection.* Nova Southeastern University, Fort Lauderdale-Davie, Florida, US, January 2005.

9. Nareli Cruz Cortes. Dissertation title: *Artificial immune system to solve problems of optimization.* The Evolutionary Computation Group at CINVESTAVIPN (EVOC-INV) 2004.

10. Giuseppe Nicosia. Dissertation Title: *Immune Algorithms for Optimization and Protein Structure Prediction.* Department of Mathematics and Computer Science, University of Catania, 2004.

11. Dr. Tom Knight. Dissertation Title: *MARIA: A Multilayered Unsupervised Machine Learning Algorithm Based on the Vertebrate Immune System*, University of Kent, Canterbury, U.K., September 2004.

12. F. Gonzalez. Dissertation Title: *A Study of Artificial Immune Systems Applied to Anomaly Detection.* Division of Computer Science, University of Memphis, Memphis, TN 38152, May 2003.

13. Anil B. Somayaji. Dissertation Title: *Operating System Stability and Security through Process Homeostasis.* Ph.D. thesis, University of New Mexico, July 2002.

14. Hossam Meshref. Dissertation Title: *Modeling Autonomous Agents' Behavior Using Neuro-Immune Networks.* Department of Electrical and Computer Engineering, Virginia Tech. 2002.

15. Jung Won Kim. Dissertation Title: *Integrating Artificial Immune Algorithms for Intrusion Detection*, Department of Computer Science, University College London, July 30, 2002.

16. E. Hart. Dissertation Title: *Immunology as a Metaphor for Computational Information Processing: Fact of Fiction*, University of Edinburgh, Scotland, U.K., 2002.

17. L. N. de Castro. Dissertation Title: *Immune Engineering: Development of Computational Tools Inspired by the Artificial Immune Systems.* (In Portuguese). DCA–FEEC/UNICAMP, Campinas/SP, Brazil, May 2001.

18. Junichi Suzuki. Dissertation Title: *Biologically-inspired Autonomous Adaptability in Communication End system: An Approach Using an Artificial Immune Network.* Keio University, 2001.

19. Lei Wang. Dissertation Title: *Immune evolutionary computation and its application.* Xidian University, 2001.

20. J. Timmis. Dissertation Title: *Artificial immune systems: A novel data analysis technique inspired by the immune network theory.* Department of Computer Science, University of Wales, Aberystwyth. Ceredigion. Wales, U.K., August 2000.

21. S. A. Hofmeyr. Dissertation Title: *An Immunological Model of Distributed Detection and its Application to Computer Security.* University of New Mexico, 1999.

22. M. Oprea. Dissertation Title: *Antibody Repertoires and Pathogen Recognition: The role of germline diversity and somatic hypermutation.* University of New Mexico. Albuquerque, NM. 1999.

23. D. J. Smith. Dissertation Title: *The Cross-Reactive Immune Response: Analysis, Modeling, and Application to Vaccine Design.* University of New Mexico, NM. 1997.
24. J. Carneiro. Dissertation Title: *Towards a comprehensive view of the immune system.* University of Porto. Portugal, 1997.
25. R. Hightower. Dissertation Title: *Computational aspect of antibody gene families.* University of New Mexico, Albuquerque, NM. 1996.
26. V. Detours. Dissertation Title: *Modeles formels de la selection des cellules B et T.,* University Paris 6, France, 1996.

Masters Thesis

1. Oladipo Lawal. Masters Thesis: *Investigation of Novel Mutation Mechanisms for Immune: Inspired Optimisation Algorithms.* School of Computing, Napier University, 2007.
2. Nrupal Choudary Prattipati. Masters Thesis: *Improvement and Evaluation of an immune-based email classification system.* School of Computing, Napier University, 2007.
3. Terri Oda. Masters Thesis: *A Spam-Detecting Artificial Immune System.* Ottawa-Charleton Institute for Computer Science, School of Computer Science, Carleton University, 2005.
4. Sankalp Balachandran. Masters Thesis: *Multi-shaped Detector generation using Real-valued representation for Anomaly Detection.* University of Memphis, Memphis, TN, US, December 2005.
5. Joseph M. Shapiro. Masters Thesis: *An Evolutionary Algorithm to Generate Ellipsoid Detectors for Negative Selection.* Air Force Institute of Technology. Wright-Patterson Air Force Base, Ohio, USA. March, 2005.
6. X. Wang. Masters Thesis: *Artificial Immune Optimization and Its Application in Industrial Electronics.* Institute of Intelligent Power Electronics, Department of Electrical and Communications Engineering, Helsinki University of Technology, 2005.
7. Amanda Marie Whitbrook. Masters Thesis: *An idiotypic immune network for mobile robot control.* School of Computer Science and Information Technology, University of Nottingham, 2005.
8. Bashar Barrishi. Masters Thesis: *Modeling the artificial immune system to the human immune system with the use of agents.* Oklahoma State University, 2004.
9. Johnny Kelsey. Masters Thesis: *An Immune Inspired Algorithm for Function Optimization.* MSc. 2004.
10. Jeong Sik Jang. Masters Thesis: *An Empirical Investigation into an Artificial Immune System for Email Classification AISEC.* 2004.
11. Lingjun Meng. Masters Thesis: *Artificial Immune System for Knowledge Discovery.* Leiden Institute of Advanced Computer Science (LIACS), Leiden University, 2004.
12. Jos_Daniel Dias Pacheco. Masters Thesis: *Computational Power of Killers and Helpers in the Immune System.* Universidade de Lisboa (Lisbon University), 2004.
13. Nyrki Rantonen. Masters Thesis: *An Artificial Immune System for Document Classification.* 2004.
14. Alex Kilgour. Masters Thesis: *Developing a Practicle Artificial Immune System for Email Classification.* 2004

15. Mark A. Esslinger. Masters Thesis: *An Artificial Immune System Strategy for Robust Chemical Spectra Classification via Distributed Heterogeneous Sensors.* Air Force Institute of Technology, Air University, 2003.

16. Alexander Jakobus Graaff. Masters Thesis: *The artificial immune system with evolved lymphocytes.* Faculty of Engineering, Built Environment and Information Technology, University of Pretoria, 2003.

17. Julie Greensmith. Masters Thesis: *New Frontiers For An Artificial Immune System.* University of Leeds, 2003.

18. Christopher C. Lord. Masters Thesis: *An Emergent Model of Immune Cognition 2003.* Carnegie Mellon University.

19. Kathy Jean Matthews. Masters Thesis: *Immunotronics: Self-repairing finite state machines 2003.* University of West England.

20. Tom Morrison. Masters Thesis: *Similarity Measure Building for Website Recommendation within an Artificial Immune System.* School of Computer Science, University of Nottingham, 2003.

21. Larissa A. O'Brien, Masters Thesis: *Using Sequence Analysis to Perform Application-Based Anomaly Detection Within an Artificial Immune System Framework.* Air Force Institute of Technology, Air University, 2003.

22. Camilla Edmonds. Masters Thesis: *Artificial Immune Networks for Function Optimisation MSc.* 2003

23. John L. Bebo. Masters Thesis: *Using Relational Schemata in a Computer Immune System to Detect Multiple-Packet Network Intrusions.* Air Force Institute of Technology, Air University, 2002.

24. Martin Thorsen Ranang. Masters Thesis: *An Artificial Immune System Approach to Preserving Security in Computer Networks.* Norwegian University of Science and Technology (NTNU), Trondheim, Norway, June 2002.

25. Lars Olsson. Masters Thesis: *Anomaly Detection Using Self/Nonself Discrimination.* Evolutionary and Adaptive Systems, The University of Sussex, 2002.

26. N. S. Majumdar. Masters Thesis: *Anomaly Detection in Single and Multidimensional datasets using Artificial Immune Systems.* Division of Computer Science, Department of Mathematical Sciences. University of Memphis. Memphis, TN. May 2002.

27. Kathia Regina Lemos Juca. Masters Thesis: *An Approach for Intrusion Detection with Immune System.* Santa Catarina Federal University, 2001.

28. Andrew B. Watkins. Masters Thesis: *AIRS: A resource limited artificial immune classifier.* Mississippi State University, 2001.

29. P. D. Williams Warthog. Masters Thesis: *Towards an artificial Immune System for detecting 'low and slow' information system attacks,* AFIT/GCE/ENG/01M-15, Air Force Institute of Technology, WPAFB, OH. March 2001.

30. Paul K. Harmer. Masters Thesis: *A Distributed Agent Architecture for a Computer Virus Immune System.* Air Force Institute of Technology, Air University, 2000.

31. Daniel Stow. Masters Thesis: *Towards an immunological approach to network management: learning, memory and cross-reactivity in an artificial immune system.* Evolutionary and Adaptive Systems, The University of Sussex, 2000.

32. Kelley J. Cardinale and Hugh M. O'Donnell. Masters Thesis: *A Constructive Induction Approach to Computer Immunology.* Air Force Institute of Technology, Air University, 1999.

Links to AIS-Related Web Sites (Last Access Date December 30, 2007)

■ People

- ❏ Uwe Aickelin: http://www.cs.nott.ac.uk/~uxa/
- ❏ Jason Brownlee: http://www.ict.swin.edu.au/personal/jbrownlee/
- ❏ D. Dasgupta: http://www.msci.memphis.edu/~dasgupta
- ❏ P. D'haeseleer: http://www-cmls.llnl.gov/?url=about_cmls-scientific_staff-dhaeseleer_p
- ❏ S. Forrest: http://www.cs.unm.edu/~forrest
- ❏ Fabio A. González: http://dis.unal.edu.co/~fgonza/
- ❏ P. Hajela: http://www.rpi.edu/~hajela
- ❏ E. Hart: http://www.dcs.napier.ac.uk/~emmah/
- ❏ S. A. Hofmeyr: http://www.cs.unm.edu/~steveah
- ❏ G.Nicosia: http://www.dmi.unict.it/~nicosia/intro.html
- ❏ N. I. Nikolaev: http://homepages.gold.ac.uk/nikolaev/
- ❏ F. Nino: http://dis.unal.edu.co/~lfnino
- ❏ L. Nunes de Castro: http://www.dca.fee.unicamp.br/~lnunes
- ❏ Mihaela Oprea, http://www.santafe.edu/~mihaela
- ❏ S. Perelson: http://www.t10.lanl.gov/asp/
- ❏ L. Segel: http://www.wisdom.weizmann.ac.il/~/NoMoreUsers/lee/
- ❏ D. J. Smith: http://www.santafe.edu/~dsmith
- ❏ S. Thayer: http://www.ri.cmu.edu/people/thayer_scott.html
- ❏ J. Timmis: http://www-users.cs.york.ac.uk/jtimmis/
- ❏ F. J. Von Zuben: http://www.dca.fee.unicamp.br/~vonzuben
- ❏ Y. Watanabe: http://www.nsc.nagoya-cu.ac.jp/profile/watanabey-e.html

■ Organizations

- ❏ CytoCom Network: http://www.csc.liv.ac.uk/~cytocom/index.html
- ❏ IBM Antivirus Research: http://www.research.ibm.com/antivirus/
- ❏ ISYS Project: http://www.aber.ac.uk/~dcswww/ISYS
- ❏ Primary Response: http://www.sanasecurity.com/

AIS-Related Events

■ 2008 Events

- ❏ The Seventh International Conference on Simulated Evolution And Learning (SEAL'08) 7–10 December 2008 Melbourne, Australia.
- ❏ The 12th International Conference on Knowledge-Based & Intelligent Information & Engineering Systems (KES2008), 3–6 September 2008, Zagreb, Croatia.
- ❏ The 4th International Conference on Natural Computation (ICNC'08) and the 5th international Conference on Fuzzy Systems and Knowledge Discovery (FSKD'08), 25th–27th August, 2008, jointly held in Jinan, China.

- ❏ 7th International Conference on Artificial Immune Systems (ICARIS 2008), 10th–13th August, Phuket, Thailand.
- ❏ The Twenty-Third AAAI Conference on Artificial Intelligence, July 13th–17th, 2008 Chicago, Illinois.
- ❏ A recombination of 17th International Conference on Genetic Algorithms (ICGA) and the 13th Annual Genetic Programming Conference (GP), 12th–16th July, 2008, Atlanta, Georgia, USA.
- ❏ International Conference on Artificial Intelligence and Pattern Recognition, 7th –10th of July 2008 in Orlando, FL, USA.
- ❏ IEEE Congress on Evolutionary Computation (CEC), June 1–6, 2008 in Hong Kong.
- ❏ 16th International Conference Intelligent Information Systems (IIS), Zakopane, Poland, June 16–18, 2008.
- ❏ The IASTED International Conference on Artificial Intelligence and Applications (AIA), Innsbruck, Austria, February 11–13, 2008.

■ *2007 Events*

- ❏ 3rd Indian International Conference on Artificial Intelligence (IICAI) 17–19 December, 2007 in Pune, India.
- ❏ 19th IEEE International Conference on Tools with Artificial Intelligence (ICTAI) 29–31 October, 2007 in Patras, Greece.
- ❏ 2nd International Symposium on Intelligence Computation and Applications (ISICA) 21–23 September, 2007 in Wuhan, China.
- ❏ IEEE Congress on Evolutionary Computation (CEC), 25–28 September, 2007 in Singapore.
- ❏ 6th International Conference on Artificial Immune Systems (ICARIS), 26–29 August, 2007 in Santos/SP, Brazil.
- ❏ The 3rd International Conference on Natural Computation (ICNC) and the 4th international Conference on Fuzzy Systems and Knowledge Discovery (FSKD), 24th–27th August, 2007, jointly held in Haikou, China.
- ❏ World Conference of STRESS, HANS SELYE 1907–2007, 23–26 August, Budapest in Hungary.
- ❏ The Twenty-Second AAAI Conference on Artificial Intelligence, July 22–26, 2007 in Vancouver, British Columbia, Canada.
- ❏ 7th Symposium on Abstraction, Reformulation, and Approximation (SARA), 18–21 July, 2007 in Vancouver, Canada.
- ❏ International Conference on Artificial Intelligence and Pattern Recognition, 9–12 of July 2007 in Orlando, FL, USA.
- ❏ The Third IASTED International Conference on Computational Intelligence, 2–4 July 2007 in Banff, Alberta, Canada.
- ❏ A recombination of 16th International Conference on Genetic Algorithms (ICGA) and the 12th Annual Genetic Programming Conference (GP) (GECCO'06), 7–11 July, 2007, London, England.
- ❏ The First IEEE Symposium on Foundations of Computational Intelligence (FOCI) 1–5 April, 2007 Hawaii, USA.

❑ The IASTED International Conference on Artificial Intelligence and Applications (AIA), February 12–14, 2007 in Innsbruck, Austria 2007.

❑ Twentieth International Joint Conference on Artificial Intelligence, 6–12 January, 2007 in Hyderabad, India.

■ *Previous Events*

❑ Tenth International Conference on Knowledge-Based & Intelligent Information & Engineering Systems (KES2006), 9–11 October 2006, Bournemouth International Conference Centre.

❑ Advisory board Member, The 2nd International Conference on Natural Computation (ICNC) and the 3rd International Conference on Fuzzy Systems and Knowledge Discovery (FSKD'06), 24–28 September 2006 will be jointly held in Xi'an, China. ICNC'06-FSKD'06.

❑ Fifth International Conference on Artificial Immune Systems (ICARIS), 4th–6th September 2006, Instituto Gulbenkian de Ciência, Oeiras, Portugal.

❑ Fifth Mexican International Conference on Artificial Intelligence, September 2006. Mexico.

❑ The 14th Annual International conference on Intelligent Systems for Molecular Biology (ISMB 2006), 6–10 August 2006, Fortaleza, Brazil.

❑ The IASTED International Conference on Computational Intelligence (CI) 17–19 July, 2006. Calgary, Alberta, Canada.

❑ A Special Session on Recent Developments in Artificial Immune Systems (IEEE World Congress on Computational Intelligence), 16–21 July 2006, Sheraton Vancouver Wall Centre Hotel, Vancouver, Canada.

❑ Artificial Immune Systems at Genetic and Evolutionary Computation Conference (GECCO'06), 8–12 July, 2006, Seattle, W.A., USA.

❑ Workshop on Artificial Immune Systems and Immune System Modelling (AISB'06: Adaptation in Artificial and Biological Systems), 4th April 2006, University of Bristol, Bristol, England.

❑ International Conference on Natural Intelligence, ICNI 2006, 24–26 February, 2006 Czech Republic, Prague.

❑ Second Indian International Conference on Artificial Intelligence (IICAI), 20th–22nd December, 2005, Pune, India.

❑ Fourth Mexican International Conference on Artificial Intelligence, 14th–18th November, 2005. Monterrey, Nuevo Leon. Mexico.

❑ A Special Session on "Immunity-Based Systems" under Information Sciences Sessions at SAE World Aerospace Congress, 4th–6th October, 2005. Grapevine, Texas, USA.

❑ Ninth International Conference on Knowledge-Based Intelligent Information & Engineering Systems (KES), 14–16 September, 2005, Melbourne, Australia.

❑ International Conference on Evolvable Systems (ICES), 12th–14th September, 2005, Sitges, Barcelona, Spain.

❑ Recent Development in Artificial Immune Systems at IEEE Congress on Evolutionary Computation (CEC), 2nd–5th September, 2005, Edinburgh, U.K.

- ❑ Sixth International Workshop on Information Processing in Cells and Tissues (IPCAT), August 30–September 1, 2005, St William's College, York, United Kingdom.
- ❑ International Conference on Natural Computation (ICNC), 27th–29th August, 2005, Changsha, China.
- ❑ International Conference on Intelligent Computing (ICIC), 23rd–26th August, 2005, Hefei, China.
- ❑ 4th International Conference on Artificial Immune Systems (ICARIS), 14th–17th August, 2005, Banff, Alberta, Canada.
- ❑ International Joint Conference on Neural Networks, 31st July–4th August, 2005, Montréal, Québec, Canada.
- ❑ Second Multidisciplinary International Conference on Scheduling: Theory and Applications (MISTA), 18th–21st July 2005, New York, USA.
- ❑ The IASTED International Conference on Computational Intelligence, 4th–6th July, 2005, Calgary, Alberta, Canada.
- ❑ Artificial Immune Systems at Genetic and Evolutionary Computation Conference (GECCO), 25th–29th June, 2005, Washington, D.C., USA.
- ❑ New Trends in Intelligent Information Processing and Web Mining (IIPWM) INTELLIGENT INFORMATION SYSTEMS 2005, IIS'05 Gdansk, Poland, June 13–16, 2005.
- ❑ International Workshop on Natural and Artificial Immune Systems (NAIS), 9th–10th June, 2005, Vietri sul Mare, Salerno, Italy.
- ❑ The Fourth IEEE International Workshop on Soft Computing as Transdisciplinary Science and Technology, 25–27 May 2005, Muroran, Japan.
- ❑ 3rd International Conference on Artificial Immune Systems, 13–16 September, 2004, Catania, Italy.
- ❑ International Workshop on "Computational Intelligence Applied to Tutoring Systems," August 30 to September 03, 2004, Maceió, Brazil.
- ❑ Special Track on Artificial Immune Systems at Genetic and Evolutionary Computation Conference (GECCO), June 26–30, 2004. Seattle, Washington, USA.
- ❑ Special Session on Artificial Immune Systems at the Congress on Evolutionary Computation (CEC), June 20–23, 2004, Portland, Oregon, USA.
- ❑ Tutorial on Immunological Computation at Mexican International Conference on Artificial Intelligence (MICAI), April 26–30, 2004, Mexico City, Mexico.
- ❑ AISB 2004 Symposium on The Immune System and Cognition (ImmCog-2004), 30th–31st March, 2004, Leeds, U.K.
- ❑ Special Session on Artificial Immune Systems at the Congress on Evolutionary Computation (CEC), December 8–12, 2003, Canberra, Australia.
- ❑ Special Session on Immunity-Based Systems at Seventh International Conference on Knowledge-Based Intelligent Information & Engineering Systems (KES), September 3–5, 2003, University of Oxford, U.K. http://www.kesinternational.org/kes 2003/ http://web.comlab.ox.ac.uk/oucl/conferences/kes2003/Invited_Sessions.html
- ❑ Second International Conference on Artificial Immune Systems (ICARIS), September 1–3, 2003, Napier University, Edinburgh, U.K.
- ❑ Tutorial on Artificial Immune Systems at First Multidisciplinary International Conference on Scheduling: Theory and Applications (MISTA), 12 August 2003, The University of Nottingham, U.K.

❑ Tutorial on Immunological Computation at International Joint Conference on Artificial Intelligence (IJCAI), 10 August 2003, Acapulco, Mexico.

❑ Special Track on Artificial Immune Systems at Genetic and Evolutionary Computation Conference (GECCO), July 12–16, 2003, Chicago, USA.

❑ 9th International Conference on Neural Information Processing, 4th Asia-Pacific Conference on Simulated Evolution and Learning, 2002 International Conference on Fuzzy Systems and Knowledge Discovery, November 18–22, 2002, Singapore. http://www.ntu.edu.sg/home/nef/

❑ Fifth International Conference on Cellular Automata for Research and Industry, October 9–11, 2002, Switzerland. This conference invites papers on immune systems as well. http://cui.unige.ch/acri2002/

❑ IEEE 2002 Systems, Man and Cybernetics conference, October 6–9 Tunisia. http://smc02.ec-lille.fr/home.html

❑ KES'2002 Special Session on Immunity-Based Systems held as part of 6th International Conference on Knowledge-Based Intelligent Information Engineering Systems, 16–18 September 2002, Podere d'Ombriano,Crema,Italy. http://www.dc.fi.udc.es/lidia/kes2002.html

❑ 1st International Conference on Artificial Immune Systems (ICARIS-2002) University of Kent, September 9–11, 2002, Canterbury. http://www.aber.ac.uk/icaris-2002/icaris-2002.htm

❑ Special track on Artificial Immune Systems held at the 2002 Congress on Evolutionary Computation as part of the 2002 IEEE World Congress on Computational Intelligence, May 12–17, 2002, Honolulu, HI. http://www.wcci2002.org/

❑ Congress On Evolutionary Computation, (CEC 2001): http://cec2001.kaist.ac.kr/

❑ Genetic and Evolutionary Computation Conference (GECCO' 2001): http://gal4.ge.uiuc.edu:8080/GECCO-2001/

❑ IEEE International Conference on Systems, Man, and Cybernetics '97, Special Track on Artificial Immune systems: http://www.msci.memphis.edu/~dasgupta/accepted-papers.html

❑ IEEE International Conference on Systems, Man, and Cybernetics '98, Special Track on Artificial Immune systems: http://www.msci.memphis.edu/~dasgupta/smc98-AIS-list.html

❑ International Workshop on Information Processing in Cells and Tissues (IPCAT' 2001): http://ipcat.etro.vub.ac.be/IPCAT2001/welcome.html

Journal Articles, Conference Papers and Technical Reports

A

1. F. Abbattista, G. Di Danto, G. Di Gioia and M. Fanelli. An associative memory based on the immune networks. In the proceedings of the International Conference on Neural Networks, 1996.

2. Adnan Acan. Clonal Selection Algorithm with Operator Multiplicity. In the proceedings of Congress on Evolutionary Computation (CEC). Portland, Oregon USA, June 19–23, 2004.

3. Aickelin Uwe and Greensmith Julie (2007): 'Sensing Danger: Innate Immunology for Intrusion Detection', Elsevier Information Security Technical Reports, pp, doi: 10.1016/j.istr.2007.10.003, Abstract [http://www.cs.nott.ac.uk/%7Euxa/papers/07eistr_danger.html], Paper [http://www.cs.nott.ac.uk/%7Euxa/papers/07eistr_danger.pdf].

4. U. Aickelin, J. Greensmith and J. Twycross. Immune System Approaches to Intrusion Detection—A Review (Conceptual Paper). Published in the proceedings of the Third International Conference on Artificial Immune Systems, Catania, Italy. September 13–16, 2004.

5. Uwe Aickelin, Peter Bentley, Steve Cayzer, Jungwon Kim and Julie McLeod. Danger Theory: The Link between AIS and IDS. Published in the Proceeding of Second International Conference on Artificial Immune Systems (ICARIS), September 1–3, 2003, Napier University, Edinburgh, U.K.

6. U. Aickelin & S. Cayzer. The Danger Theory and Its Application to Artificial Immune Systems. Published in the proceedings of 1st International Conference on Artificial Immune Systems (ICARIS), University of Kent at Canterbury, U.K., September 9–11, 2002.

7. H. Aisu and H. Mizutani. Immunity-based learning—Integration of distributed search and constraint relaxation. Presented at ICMAS Workshop on Immunity-Based Systems, December 10, 1996.

8. Oscar Alonso, Fabio A. Gonzalez, Fernando Niño, Juan Galeano. Search and Optimization: A Solution Concept for Artificial Immune Networks: A Coevolutionary Perspective. In the proceedings of 6th international conference on Artificial Immune systems, 26th–29th August, 2007 in Santos/SP, Brazil.

9. O. M. Alonso, F. Nino and M. Velez. A Robust Immune Based Approach To the Iterated Prisoner's Dilemma (Conceptual Paper). Published in the proceedings of the Third International Conference on Artificial Immune Systems (ICARIS), Catania, Italy. September 13–16, 2004.

10. Jorge Amaral, Jose Amaral and Ricardo Tanscheit. An Immune Fault Detection System for Analog Circuits with Automatic Detector Generation. Published in the proceedings of IEEE World Congress on Computational Intelligence in Congress on Evolutionary Computation, Vancouver, Canada, July 16–21, 2006.

11. Anchor, Zydallis, Gunsch & Lamont. Extending the Computer Defense Immune System: Network Intrusion Detection with a Multiobjective Evolutionary Programming Approach. Published in the proceedings of 1st International Conference on Artificial Immune Systems (ICARIS), University of Kent at Canterbury, U.K., September 9–11, 2002.

12. K. P. Anchor, P. D. Williams, G. H. Gunsch and G. B. Lamont. The Computer Defense Immune System: Current and Future Research in Intrusion Detection. Published in the proceedings of the special sessions on artificial immune systems in Congress on Evolutionary Computation, IEEE World Congress on Computational Intelligence, Honolulu, Hawaii, May 2002.

13. Shin Ando. Artificial Immune System for Classification of Gene Expression Data. Published in the proceedings of the Genetic and Evolutionary Computation Conference (GECCO), Chicago, IL, USA, July 2003.

14. Paul S. Andrews and Jon Timmis. A Computational Model of Degeneracy in a Lymph Node. Published in the proceedings of the 5th International Conference on Artificial Immune Systems (ICARIS), Portugal, 4–6 September, 2006.

15. Paul S. Andrews and J Timmis. Inspiration for the Next Generation of Artificial Immune Systems. Published in the Proceedings of ICARIS, 4th International Conference on Artificial Immune Systems, Banff, Canada, 2005.

16. Secker Andrew, Alex A. Freitas, Jon Timmis. AISEC: an Artificial Immune System for E-mail Classification. Published in the proceedings of the IEEE Congress on Evolutionary Computation, Canberra, Australia, December 8–12, 2003.

17. Nikolaos D. Atreas, Costas G. Karanikas and Alexander Tarakanov. Signal Processing by an Immune Type Tree Transform. Published in the Proceeding of Second International Conference on Artificial Immune Systems (ICARIS), September 1–3, 2003, Napier University, Edinburgh, U.K.

18. G. H. Anthes. Future Watch: Immune Computer Systems. Computer World Magazine, December 9, 2002. Link [http://www.computerworld.com/securitytopics/security/story/0,10801,76412,00.html].

19. M. Araujo, J. Aguilar, H. Aponte. Fault detection system in gas lift well based on Artificial Immune System. Published in the proceedings of the International Joint Conference on Neural Networks. pp. 1673–1677, Vol. 3, No.3, July 20–24, 2003. Link [http://ieeexplore.ieee.org/xpl/tocresult.jsp?isNumber=27485&page=0].

20. M. Villalobos-Arias, C.A. Coello Coello and O. Hernandez-Lerma. Convergence Analysis of a multi objective Artificial Immune System Algorithm. Published in the proceedings of the Third International Conference on Artificial Immune Systems (ICARIS), Catania, Italy. September 13–16, 2004.

21. Veysel Aslantas, Saban Ozer, Serkan Ozturk. General Applications: A Novel Clonal Selection Algorithm Based Fragile Watermarking Method. In the proceedings of 6th international conference on Artificial Immune systems, 26th–29th August, 2007 in Santos/SP, Brazil.

22. R. R. F. Attux, M. B. Loiola, R. Suyama, L. N. de Castro, F. J. Von Zuben & J. M. T. Romano (2003), "Blind Search for Optimal Wiener Solutions Using an Artificial Immune Network Model," EURASIP Journal of Applied Signal Processing, Special Issue on Genetic and Evolutionary Computation for Signal Processing and Image Analysis (in print).

23. M Ayara, J Timmis, Rogério de Lemos, Forrest S. Immunising Automated Teller Machines. Published in the Proceedings of ICARIS, 4th International Conference on Artificial Immune Systems, Banff, Canada, 2005.

24. M Ayara, J Timmis, de Lemos, de Castro & Duncan. Negative Selection: How to Generate Detectors. Published in the proceedings of 1st International Conference on Artificial Immune Systems (ICARIS), University of Kent at Canterbury, U.K., September 9–11, 2002.

B

25. M. Bakhouya, J. Gaber, A. Koukam. Immune-based middleware for large-scale network. Published in the proceedings of Local Computer Networks (LCN), 27th Annual IEEE Conference pp. 230–231. November 6–8, 2002.

26. M. Bakhouya, J. Gaber, A. Koukam. A Middleware for large Scale Networks Inspired by the Immune System. International Parallel and Distributed Processing Symposium: IPDPS workshop, Fort Lauderdale, Florida. April 15–19, 2002.

27. S. Balachandran, Dipankar Dasgupta, Fernando Nino and Deon Garrett. A framework for evolving multi-shaped detectors in negative selection. To appear in the proceedings of the First IEEE Symposium on Foundations of Computational Intelligence (FOCI) 1–5 April 2007, Honolulu, Hawaii, USA.

28. P. Ballet, J. Abgrall and V. Rodin. Simulation of thrombin generation during plasmatic coagulation and primary hemostasis. Published in the proceedings of IEEE International Conference on Systems, Man and Cybernetics (SMC), Nashville, October 8–11, 2000.

29. P. Ballet and V. Rodin. Immune Mechanisms to Regulate Multi-Agents Systems. Published in the proceedings of the International Conference Genetic and Evolutionary Computation (GECCO), Las Vegas, Nevada, USA, July 8, 2000.

30. P. Ballet, J. Tisseau and F. Harrouet. A Multi-agent system to model a human humoral response. Published in the proceedings of the IEEE International Conference on Systems, Man, and Cybernetics, Orlando, Florida, October 13, 1997.

31. J. Balthrop, F. Esponda, S. Forrest and M. R. Glickman. Coverage and Generalization in an Artificial Immune System (AAAA). Published in the proceedings of the International Conference Genetic and Evolutionary Computation (GECCO), New York, July 9–13, 2002.

32. J. Balthrop, S. Forrest and M. R. Glickman. Revisting LISYS: Parameters and Normal Behavior. Published in the proceedings of the special sessions on artificial immune systems in the 2002 Congress on Evolutionary Computation, IEEE World Congress on Computational Intelligence, Honolulu, Hawaii. 2002.

33. Catherine Beauchemin, Stephanie Forrest and Frederick T. Koster. Modelling Influenza Viral Dynamics in Tissue. Published in the proceedings of the 5th International Conference on Artificial Immune Systems (ICARIS), Portugal, 4–6 September, 2006.

34. Esma Bendiab, Souham Meshoul Published in the Proceedings of Second International Conference on Artificial Immune Systems (ICARIS), September 1–3, 2003, Napier University, Edinburgh, U.K.

35. E. Benjamini, G. Sunshine and S. Leskowitz. Immunology: A Short course. Wiley-Liss, Inc. New York, third edition, 1996.

36. Peter J Bentley, Greensmith J, Ujjin S. Two Ways to Grow Tissue for Artificial Immune Systems. Published in the proceedings of ICARIS, 4th International Conference on Artificial Immune Systems, Banff, Canada, 2005.

37. Peter J. Bentley, Gordana Novakovic, Anthony Ruto. Fugue: An Interactive Immersive Audio visualisation and Artwork Using an Artificial Immune System. Published in the proceedings of ICARIS, 4th International Conference on Artificial Immune Systems, Banff, Canada, 2005.

38. P.J. Bentley and J. Timmis. A Fractal Immune Network (Conceptual Paper). Published in the proceedings of the Third International Conference on Artificial Immune Systems (ICARIS), Catania, Italy. September 13–16, 2004.

39. Priscila C. Berbert, Leonardo J. R. Freitas Filho, Tiago A. Almeida, Márcia B. Carvalho, Akebo Yamakami. Applications and Negative Selection: Artificial Immune System to Find a Set of k-Spanning Trees with Low Costs and Distinct Topologies. In the proceedings of 6th international conference on Artificial Immune systems, 26th–29th August, 2007 in Santos/SP, Brazil.

40. Heder S. Bernardino, Helio J.C. Barbosa, Afonso C.C. Lemonge. Constraint Handling in Genetic Algorithms via Artificial Immune Systems. Published in the proceedings of the International Conference Genetic and Evolutionary Computation (GECCO) 2006.

41. Hugues Bersini. Immune System Modeling: The OO Way. Published in the proceedings of the 5th International Conference on Artificial Immune Systems (ICARIS), Portugal, 4–6 September, 2006.

42. H. Bersini. Why the First Glass of Wine Is Better Than the Seventh. Published in the proceedings of ICARIS, 4th International Conference on Artificial Immune Systems, Banff, Canada, 2005.

43. H. Bersini. Self-Assertion versus Self-Recognition: A Tribute to Francisco Varela. Published in the proceedings of 1st International Conference on Artificial Immune Systems (ICARIS), University of Kent at Canterbury, U.K., September 9th–11th, 2002.

44. H. Bersini. The Immune and Chemical Crossover. In the Special Issue on Artificial Immune Systems of the journal IEEE Transactions on Evolutionary Computation, Vol. 6, No. 3, June 2002.

45. H. Bersini. The Endogenous Double Plasticity of the Immune Network and the Inspiration to be Drawn for Engineering Artifacts. Chapter 2 in the book entitled Artificial Immune Systems and Their Applications, (D. Dasgupta, editor) Springer-Verlag, Inc., pp. 22–41, January 1999.

46. H. Bersini and V. Calenbuhr. Frustrated Chaos in Biological Networks. In Journal of Theoretical Biology, Vol. 188, No 2, pp. 187–200, 1996.

47. H. Bersini and V. Calenbuhr. Frustration Induced Chaos in a System of Coupled ODE'S. In Chaos, Soliton and Fractals, Vol. 5, No 8, pp. 1533–1549, 1995.

48. H. Bersini and F. Varela. Computing with Biological Metaphors. Chapman-Hall. Chap. The immune learning mechanisms: Reinforcement and Recruitment and their applications. pp. 166–192. 1994.

49. H. Bersini and F. Varela. The Immune Learning Mechanisms: Recruitment R-L-enforcement and their applications. In Computing with Biological Metaphors, Chapman and Hall, R. Patton (Ed.), 1993.

50. H. Bersini and F. Varela. The Immune Recruitment Mechanism: A Selective Evolutionary Strategy. Published in the proceedings of the 4th International Conference on Genetic Algorithms, R. Belew and L. Booker (Eds.), Morgan Kaufman, pp. 520–526, 1991.

51. H. Bersini and F. Varela. Hints for Adaptive Problem Solving Gleaned from Immune Network. In Parallel Problem Solving from Nature, H.P. Schwefel and H. M'hlenbein (Eds.), Publisher Springer-Verlag, pp. 343–354, 1990.

52. George B. Bezerra, Tiago V. Barra, Hamilton M. Ferreira, Helder Knidel, Leandro Nunes de Castro and Fernando J. Von Zuben. An Immunological Filter for Spam. Published in the proceedings of the 5th International Conference on Artificial Immune Systems (ICARIS), Portugal, 4–6 September, 2006.

53. George B. Bezerra, Tiago V. Barra, Leandro N. de Castro, Fernando J. Von Zuben. Adaptive Radius Immune Algorithm for Data Clustering. Published in the proceedings of ICARIS, 4th International Conference on Artificial Immune Systems, Banff, Canada, 2005.

54. G.B. Bezerra, L.N. de Castro and F.J. Von Zuben. A Hierarchical Immune Network Applied to Gene Expression Data. Published in the proceedings of the Third International Conference on Artificial Immune Systems (ICARIS), Catania, Italy. September 13–16, 2004.

55. George Barreto Bezerra and Leandro Nunes de Castro. Bioinformatics Data Analysis Using an Artificial Immune Network. Published in the proceedings of Second International Conference on Artificial Immune Systems (ICARIS), September 1–3, 2003, Napier University, Edinburgh, U.K. Jean-Yves Le Boudec and Slavisa Sarafijanovic. An Artificial Immune System Approach to Misbehavior Detection in Mobile Ad-Hoc Networks *Bio-ADIT* 2004, pp. 96–111, 29–30 Jan 2004, Lausanne, Switzerland. Link [http://www.terminodes.org/micsPublicationsDetail.php?pubno=601]

56. D. W. Bradley and A. M Tyrrell. Immunotronics: Novel Finite-State-Machine Architecture with Built-in Self-Test Using Self-Nonself Differentiation. In the Special Issue on Artificial Immune Systems of the journal IEEE Transactions on Evolutionary Computation, Vol. 6, No. 3, June 2002.

57. D. Bradley and A. M. Tyrrell. A Hardware Immune System for Benchmark State Machine Error Detection. Published in the proceedings of the special sessions on artificial immune systems in Congress on Evolutionary Computation, IEEE World Congress on Computational Intelligence, Honolulu, Hawaii, May 2002.

58. D. Bradley and A. M. Tyrrell. Hardware Fault Tolerance: An Immunological Solution. Published in the proceedings of IEEE International Conference on Systems, Man and Cybernetics (SMC), Nashville, October 8–11, 2000.

59. D. Bradley and A. M. Tyrrell. Immunotronics: Hardware Fault Tolerance Inspired by the Immune System. In J. Miller, A. Thompson, P. Thomson, and T.C. Fogarty. (Eds.) Third International Conference on Evolvable Systems (ICES), Volume 1801 of Lecture Notes in Computer Science, pp. 11–20. Springer-Verlag, April 2000.

60. P.J.C. Branco, J.A. Dente, R.V. Mendes. Using Immunology Principles for Fault Detection Industrial Electronics, IEEE Transactions, pp. 362–373, Vol. 50, No. 2. April 2003. Link [http://pbranco.ist.utl.pt/publications.htm]

61. Jason Brownlee. IIDLE: An Immunological Inspired Distributed Learning Environment for Multiple Objective and Hybrid Optimisation. Published in the proceedings of IEEE World Congress on Computational Intelligence (special session on recent development in artificial immune systems) in Congress on Evolutionary Computation, Vancouver, Canada, July 17–21, 2006.

62. C. Bruce Trapnell Jr. A Peer-to-Peer Blacklisting Strategy Inspired by Leukocyte-Endothelium Interaction. Published in the proceedings of ICARIS, 4th International Conference on Artificial Immune Systems, Banff, Canada, 2005.

63. P. Bull, A. Knowles, G. Tedesco and A. Hone. Diophantine Benchmarks for the B-Cell Algorithm. Published in the proceedings of the 5th International Conference on Artificial Immune Systems (ICARIS), Portugal, 4–6 September, 2006.

64. M. Burgess. Evaluating cfengine's immunity model of site maintenance. Published in the Proceedings of the SANE 2000 conference.

65. M. Burgess, O. College. Computer Immunology. Published in the proceedings of the Twelfth systems Administration Conferences (LISA). Boston Massachusetts, December 6–11, 1998.

C

66. M. Caetano, J. Manzolli, F. J. V. Zuben Application of an Artificial Immune System in a Compositional Timbre Design Technique. Published in the proceedings of ICARIS, 4th International Conference on Artificial Immune Systems, Banff, Canada, 2005.

67. V. Calenbuhr, F. Varela and H. Bersini. Immune Idiotypic Network. In International Journal of Bifurcation and Chaos, Vol. 6 No 9, pp. 1691–1702, 1996.

68. V. Calenbuhr, H. Bersini, J. Stewart and F. Varela. Natural Tolerance in a Simple Immune Network. In Journal of Theoretical Biology, 177, pp. 199–213, 1995.

69. V. Calenbuhr, H. Bersini, F. J. Varela and J. Stewart. The Impact of the Structure of the Connectivity Matrix on the Dynamics of a Simple Model for the Immune Network. Published in the proceedings of the First Copenhagen Symposium on Computer Simulation in Biology, Ecology and Medicine - Mosekilde, E. (Ed.), pp. 41–45, 1993.

70. Felipe Campelo, Frederico Guimaraes, Hajime Igarashi, Kota Watanabe and Jaime Ramirez. An Immune-based Algorithm for Topology Optimization. Published in the proceedings of IEEE World Congress on Computational Intelligence in Congress on Evolutionary Computation, Vancouver, Canada, July 16–21, 2006.

71. R. Canham, A.H. Jackson, A. Tyrrell. Robot Error Detection using an Artificial Immune System. Evolvable Hardware, Published in the proceedings of NASA/DoD Conference. pp. 199–207, July 9–11 2003.

72. Canham and Tyrrell. A Multilayered Immune System for Hardware Fault Tolerance within an Embyronic Array. Published in the proceedings of 1st International Conference on Artificial Immune Systems (ICARIS), University of Kent at Canterbury, U.K., September 9th–11th, 2002.

73. A. Canova, F. Freschi and M. Repetto. Hybrid method coupling AIS and zeroth order deterministic search, COMPEL: The International Journal for Computation and Mathematics in Electrical and Electronic Engineering Vol. 24 No. 3, 2005, pp. 784–795.

74. Y. Cao and D. Dasgupta. An Immunogenetic Approach in Chemical Spectrum Recognition. Chapter 36 in the edited volume Advances in Evolutionary Computing (Ghosh & Tsutsui, eds.), Springer-Verlag, Inc. January 2003.

75. K. J. Cardinale and H. M. O'Donnell. A constructive introduction approach to computer immunology. M. S. thesis. Air Force Institute of Technology. WPAFB. OH. March 1999. AFIT/GCE/ENG/99M-02.

76. J. Carneiro, J. Faro, A. Coutinho and J. Stewart. A model of the immune network with B-T cell co-operation. I-Prototypical Structures and Dynamics. J. Theor. Biol. 182, 513, 1996.

77. J. Carneiro, A. Coutinho and J. Stewart. A model of the immune network with B-T cell co-operation. II-The simulation of ontogenesis. Journal of Theoretical Biology, 182, 531, 1996.

78. J. H. Carter. The Immune System as a Model for Pattern Recognition and Classification. Journal of the American Medical Informatics Association. Vol. 7, no. 3, pp. 28–41, 2000.

79. D. R. Carvalho and A. A. Freitas. An immunological algorithm for discovering small-disjunct rules in data mining. Published in the proceedings of Genetic and Evolutionary Computation Conference GECCO, (Workshop Program.) San Francisco. California. July 7, 2001.

80. Enrico Carpaneto, Claudio Cavallero, Fabio Freschi and Maurizio Repetto. Immune Procedure for Optimal Scheduling of Complex Energy Systems. Published in the proceedings of the 5th International Conference on Artificial Immune Systems (ICARIS), Portugal, 4–6 September, 2006.

81. Pablo A. D. Castro and Fernando J. Von Zuben. Bayesian Learning of Neural Networks by Means of Artificial Immune Systems. Published in the proceedings of IEEE World Congress on Computational Intelligence in Congress on Evolutionary Computation, Vancouver, Canada, July 16–21, 2006.

82. Steve Cayzer and Julie Sullivan. Modeling danger and energy in artificial immune systems. Published in the Proceedings of the 9th annual conference on Genetic and evolutionary computation (GECCO) 2007, pp. 26–32, London, England.

83. Steve Cayzer and Jim Smith. Gene Libraries: Coverage, efficiency and diversity. Published in the proceedings of the 5th International Conference on Artificial Immune Systems (ICARIS), Portugal, 4–6 September, 2006.

84. S. Cayzer, J Smith, A.R. James Marshall, T. Kovacs. What Have Gene Libraries Done for AIS? Published in the proceedings of ICARIS, 4th International Conference on Artificial Immune Systems, Banff, Canada-2005.

85. S. Cayzer and U. Aickelin. A Recommender System based on Idiotypic Artificial Immune Networks. In the Journal Mathematical Modeling and Algorithms, in print, 2004.

86. S. Cayzer & U. Aickelin. On the Effects of Idiotypic Interactions for Recommendation Communities in Artificial Immune Systems. Published in the proceedings of 1st International Conference on Artificial Immune Systems (ICARIS), University of Kent at Canterbury, U.K., September 9th–11th, 2002.

87. S. Cayzer and U. Aickelin. A Recommender System based on the Immune Network. Published in the proceedings of the special sessions on artificial immune systems in Congress on Evolutionary Computation, IEEE World Congress on Computational Intelligence, Honolulu, Hawaii, May 2002.

88. Renato Reder Cazangi and Fernando Von Zuben. Immune Learning Classifier Networks: Evolving Nodes and Connections. Published in the proceedings of IEEE World Congress on Computational Intelligence in Congress on Evolutionary Computation, Vancouver, Canada, July 16–21, 2006.

89. F. Celada and P. E. Seiden. Modeling Immune Cognition. Published in the proceedings of the IEEE International Conference on Systems, Man, and Cybernetics, October 11–14, 1998.

90. F. Celada and P. E. Seiden. Affinity maturation and hypermutation in a simulation of the humoral immune response. European Journal of Immunology 26:1350–1358, 1996.

91. F. Celada and Philip E. Seiden. A computer model of cellular interactions in the immune system. Immunology Today, 13(2): 56–62, 1992.

92. Hyi Taek Ceong, Young-il Kim, Doheon Lee and Kwang H. Lee. Complementary Dual Detectors for Effective Classification. Published in the proceeding of Second International Conference on Artificial Immune Systems (ICARIS), September 1–3, 2003, Napier University, Edinburgh, U.K.

93. D. L. Chao, M. P. Davenport, S. Forrest, and A. S. Perelson. Modeling the impact of antigen kinetics on T-cell activation and response. *Immunology and Cell Biology*, 82(1): 55–61. 2004.

94. D. L. Chao, M. P. Davenport, S. Forrest, and A. S. Perelson. A stochastic model of cytotoxic T cell responses. *Journal of Theoretical Biology*, 228(2): 227–240. 2004.

95. Dennis L. Chao and Stephanie Forrest. Generating Biomorphs with an Aesthetic Immune System. In book "Artificial Life VIII: Published in the proceedings of the Eighth International Conference on the Simulation and Synthesis of Living Systems"89–92, MIT Press, Sydney, Australia, 2003.

96. Dennis L. Chao and Stephanie Forrest. Information immune systems, Genetic Programming and Evolvable Machines, December 2003.

97. Dennis L. Chao, M.P. Davenport, S. Forrest, A.S. Perelson. Stochastic stage-structured modeling of the Adaptive Immune System. In Bioinformatics Conference, CSB 2003. pp. 124–131. Published in the proceedings of the IEEE, August 11–14, 2003.

98. Dennis L. Chao and Stephanie Forrest, Information immune systems. Published in the proceedings of the International Conference on Artificial Immune Systems (ICARIS), University of Kent at Canterbury, U.K., pp. 132–140, September 9–11, 2002.

99. Jun Chen and Mahdi Mahfouf. A Population Adaptive Based Immune Algorithm for Solving Multi-Objective Optimization Problems. Published in the proceedings of the 5th International Conference on Artificial Immune Systems (ICARIS), Portugal, 4–6 September, 2006.

100. J. J. Chen. A Heuristic Approach to Efficient Production of Detector Sets For An Artificial Immune Algorithm-Based Bankruptcy Prediction System in Portfolio Management. Published in the proceedings of the special sessions on artificial immune systems in Congress on Evolutionary Computation, IEEE World Congress on Computational Intelligence, Honolulu, Hawaii, May 2002.

101. D. Chowdhury. Immune Networks: An Example of Complex Adaptive Systems. Chapter 5 in the book entitled Artificial Immune Systems and Their Applications, Publisher: Springer-Verlag, Inc., pp. 89–102, January 1999.

102. D. Chowdhury. Roles of intra-clonal and inter-clonal interactions in immunological memory: illustration with a toy model. Ind. J. Phys. 69B, 539, 1995.

103. D. Chowdhury, J. K. Bhattacharjee and A. Bhattacharya. Dynamics of crumpling of fluid-like amphiphilic membranes. Journal of Physics A (IOP, U.K.), Vol. 27, 257, 1994.

104. D. Chowdhury, V. Deshpande and D. Stauffer. Modeling immune network through cellular automata: a unified mechanism of immunological memory. International Journal of Modern Physics C (World Sc.), Vol. 5, 1049, 1994.

105. D. Chowdhury. A unified model of immune response. II: continuum approach. Journal of Theoretical Biology (Academic Press), Vol. 165, 135, 1993.

106. D. Chowdhury and D. Stauffer. Statistical Physics of Immune Networks. Physica A (Elsevier), Vol. 186, 61–81, 1992.

107. D. Chowdhury and D. Stauffer. Bursting of soap films. Physica A (Elsevier), Vol. 186, 237–249, 1992.

108. D. Chowdhury, M. Sahimi and D. Stauffer. A Discrete Model for Immune Surveillance, Tumor Immunity and Cancer. Journal of Theoretical Biology (Academic Press), Vol. 152, 263, 1991.

109. D. Chowdhury and D. Stauffer. Zellularautomaten in der Immunologie. (In German), Magazin Fuer Computer Technik, p. 204, February 1991.

110. D. Chowdhury and D. Stauffer. Systematics of the Models of Immune Response and Autoimmune Disease. Journal of Statistical Physics (Plenum), Vol. 59, 1019, 1990.
111. D. Chowdhury, D. Stauffer and P. V. Choudary. A Unified Discrete Model of Immune Response. Journal of Theoretical Biology (Academic Press), Vol. 145, 207, 1990.
112. D. Chowdhury and B. K. Chakrabarti. Robustness of the Network Models of Immune Response. Physica A (Elsevier), Vol. 167, 635, 1990.
113. J. S. Chun et al. A study on Comparison between immune algorithm and the other algorithms. In: ISAP'97. 588–592. 1997.
114. Carlos A. Coello Coello, Daniel Cortes Rivera and Nareli Cruz Cortes. Use of an Artificial Immune System for Job Shop Scheduling. Published in the proceeding of Second International Conference on Artificial Immune Systems (ICARIS), September 1–3, 2003, Napier University, Edinburgh, U.K.
115. Carlos A. Coello Coello & Cruz Cortes. An Approach to Solve Multiobjective Optimization Problems Based on an Artificial Immune System. Published in the proceedings of 1st International Conference on Artificial Immune Systems (ICARIS), University of Kent at Canterbury, U.K., September 9–11, 2002.
116. D. Chowdhury, P. D. Castro, G. P. Coelho, M. F. Caetano, F. J. V. Zuben. Designing Ensembles of Fuzzy Classification Systems: An Immune-Inspired Approach. Published in the proceedings of ICARIS, 4th International Conference on Artificial Immune Systems, Banff, Canada, 2005.
117. Krzysztof Ciesielski, Slawomir T. Wierzchon and Mieczyslaw A. Klopotek. An Immune Network for Contextual Text Data Clustering. Published in the proceedings of the 5th International Conference on Artificial Immune Systems (ICARIS), Portugal, 4–6 September, 2006.
118. E. Clark, A. Hone, J. Timmis. A Markov Chain Model of the B-Cell Algorithm. Published in the proceedings of ICARIS, 4th International Conference on Artificial Immune Systems, Banff, Canada, 2005.
119. C. A. Coello Coello and N. C. Cortes. A parallel implementation of the Artificial Immune System to handle Constraints in Genetic Algorithms: Preliminary Results. Published in the proceedings of the special sessions on artificial immune systems in Congress on Evolutionary Computation, IEEE World Congress on Computational Intelligence, Honolulu, Hawaii, May 2002.
120. I. R. Cohen. The cognitive paradigm and the immunological homunculus. Immunology Today, 13(12): 490–404, 1992.
121. I. R. Cohen. The cognitive principle challenges clonal selection. Immunology Today, Vol. 13, pp. 441–444, 1992.
122. I. R. Cohen. A cognitive paradigm of the immune system. Immunology Today, Vol. 13, pp. 490–494, 1992.
123. D. E. Cooke and J. E. Hunt. Recognizing Promoter Sequences Using an Artificial Immune System. Published in the proceedings Intelligent Systems in Molecular Biology (ISMB'95), Pub AAAI Press, pp. 89–97, 1995.
124. Nareli Cruz Cortés and Carlos A. Coello Coello. Multiobjective Optimization Using Ideas from the Clonal Selection Principle. Published in the proceedings of the Genetic and Evolutionary Computation Conference (GECCO), Chicago, IL, USA, LNCS 2723, p. 158 ff, July 12–16, 2003.

125. A. M. Costa, P. A. Vargas, F. J. Von Zuben and P. M. Franca. Makespan Minimization on Parallel Processors: An Immune-Based Approach. Published in the proceedings of the special sessions on artificial immune systems in Congress on Evolutionary Computation, IEEE World Congress on Computational Intelligence, Honolulu, Hawaii, May 2002.

126. A. Coutinho. Beyond clonal selection and network. Immunol. Rev. 110, 63, 1989.

127. A. Coutinho. The self non-self discrimination and the nature and acquisition of the antibody repertoire. Annals of Immunology. (Inst. Past.) 131D. 1980.

128. G. Cserey, W. Porod and T. Roska. An Artificial Immune System based Virtual Analysis Model and its Real-Time Terrain Surveillance Application. Published in the proceedings of the Third International Conference on Artificial Immune Systems (ICARIS), Catania, Italy. September 13–16, 2004.

129. G. Cserey and T. Roska. Artificial immune systems based novelty detection with CNN-UM. To appear in the proceedings of the First IEEE Symposium on Foundations of Computational Intelligence (FOCI) 1–5 April 2007, Honolulu, Hawaii, USA.

130. V. Cutello, D. Lee, G. Nicosia, M. Pavone and I. Prizzi Aligning Multiple Protein Sequences by Hybrid Clonal Selection Algorithm with Insert-Remove-Gaps and BlockShuffling Operators. Published in the proceedings of the 5th International Conference on Artificial Immune Systems (ICARIS), Portugal, 4–6 September, 2006.

131. V. Cutello, G. Narzisi, G. Nicosia, Pavone M. Clonal Selection Algorithms: A Comparative Case Study Using Effective Mutation Potentials. Published in the proceedings of ICARIS, 4th International Conference on Artificial Immune Systems, Banff, Canada, 2005.

132. V. Cutello, G. Nicosia, G. Narzisi, A.M. Anile, S. Spinella. Lipschitzian pattern search and Immunological Algorithm with quasi-Newton method for the Protein Folding Problem: An innovative multistage approach. Published in the proceedings of International Workshop on Natural and Artificial Immune Systems (NAIS) Vietri sul Mare, Salerno, Italy, June 9–10, 2005.

133. V. Cutello, G. Nicosia. A Clonal Selection Algorithm for Coloring, Hitting Set and Satisfiability Problems. Published in the proceedings of International Workshop on Natural and Artificial Immune Systems (NAIS) Vietri sul Mare, Salerno, Italy, June 9–10, 2005.

134. Vincenzo Cutello, Giuseppe Nicosia, and Mario Pavone. A Hybrid Immune Algorithm with Information Gain for the Graph Coloring Problem. Published in the proceedings of the Genetic and Evolutionary Computation Conference (GECCO), Chicago, IL, USA, July 12–16, 2003. LNCS 2723, p. 171 ff.

135. V. Cutello, G. Nicosia and M. Pavone. Exploring the Capability of Immune Algorithms: A Characterization of Hypermutation Operators. Published in the proceedings of the Third International Conference on Artificial Immune Systems (ICARIS), Catania, Italy. September 13–16, 2004.

D

136. Dipankar Dasgupta, Immuno-Inspired Autonomic System for Cyber Defense. Published in the Journal: Information Security Technical Report, Elsevier Ltd, Volume 12, issue 4, December 2007. (Same as Computer Science Technical Report, May, 2004.)

137. Dipankar Dasgupta. Advances in Artificial Immune Systems Published in IEEE Computational Intelligence Magazine November 2006.

138. D. Dasgupta and F. Gonzalez. Artificial Immune Systems in Intrusion Detection, Chapter 7 in the book 'Enhancing Computer Security with Smart Technology' (Editor V. Rao Vemuri), pages 165–208, Auerbach Publications, November 2005.

139. D. Dasgupta, S. Yu and N. Majumdar. MILA. Multi-Level Immune Learning Algorithm and its application to Anomaly Detection. In the Soft Computing Journal, Vol. 9, No. 3, pp. 172–184 (Online Publication was in December 2003) March 2005.

140. D. Dasgupta, K. Krishnakumar, D. Wong and M. Berry. Negative Selection Algorithm for Aircraft Fault Detection. Published in the proceedings of the 3rd International Conference on Artificial Immune Systems, Italy, September 13–16, 2004.

141. D. Dasgupta, K. KrishnaKumar, D. Wong and M. Berry. Negative Selection Algorithm for Aircraft Fault Detection. Published in the proceedings of the Third International Conference on Artificial Immune Systems (ICARIS), Catania, Italy. September 13–16, 2004.

142. D. Dasgupta and J. Zhou. Reviewing the development of AIS in last five years. Published at the 2003 IEEE Congress on Evolutionary Computation, Canberra, Australia, December 8th–12th, 2003.

143. D. Dasgupta, S. Yu and N. S. Majumdar. 'MILA—Multilevel Immune Learning Algorithm'. Published in the proceedings of the Genetic and Evolutionary Computation Conference (GECCO), Chicago, July 12–16, 2003. LNCS 2723, p. 183 ff.

144. D. Dasgupta and F. Gonzalez. An Immunity-Based Technique to Characterize Intrusions in Computer Networks. In the Special Issue on Artificial Immune Systems of the journal IEEE Transactions on Evolutionary Computation, Vol. 6, No. 3, June 2002.

145. D. Dasgupta and N. S. Majumdar. Anomaly Detection in Multidimensional Data using Negative Selection Algorithm. Published in the proceedings of the special sessions on artificial immune systems in Congress on Evolutionary Computation, IEEE World Congress on Computational Intelligence, Honolulu, Hawaii, May 2002.

146. D. Dasgupta, N. Majumdar and S. Yu. Multi-Level Immune Learning Algorithm: Preliminary Results. Technical Report CS-02-003, May 2002.

147. D. Dasgupta and F. Gonzalez. An Immunogenetic Approach to Intrusion Detection, CS Technical Report (No. CS-01-001), University of Memphis, May 2001.

148. D. Dasgupta and F. Nino. A Comparison of Negative and Positive Selection Algorithms in Novel Pattern Detection. Published in the proceedings of the IEEE International Conference on Systems, Man and Cybernetics (SMC), Nashville, October 8–11, 2000.

149. D. Dasgupta. An Immune Agent Architecture for Intrusion Detection. Published in the proceedings of Genetic and Evolutionary Computation Conference (GECCO), Las Vegas, Nevada, USA, July 8, 2000.

150. D. Dasgupta and M. Krishnan. Role of Germinal Centers: From a Computational Viewpoint. Published in the proceedings of Genetic and Evolutionary Computation Conference (GECCO), Las Vegas, Nevada, USA, July 8, 2000.

151. D. Dasgupta, Y. Cao and C. Yang. An Immunogenetic Approach to Spectra Recognition. Published in the proceedings of the International Conference Genetic and Evolutionary Computation (GECCO), July 13–17, 1999.

152. D. Dasgupta and S. Forrest. Artificial Immune Systems in Industrial Applications. Published in the proceedings of the Second International Conference on Intelligent Processing and Manufacturing of Materials (IPMM), Honolulu, July 10–15, 1999.

153. D. Dasgupta and S. Forrest. An Anomaly Detection Algorithm Inspired by the Immune System. Chapter 14 in the book entitled Artificial Immune Systems and Their Applications, Publisher: Springer-Verlag, Inc., pp. 262–277, January 1999.

154. D. Dasgupta. An Overview of Artificial Immune Systems and Their Applications. Chapter 1 in the book entitled Artificial Immune Systems and Their Applications, Publisher: Springer-Verlag, Inc., pp. 3–23, January 1999.

155. D. Dasgupta. Information Processing Mechanisms of the Immune System, A chapter in the book, "New Ideas in Optimization". McGraw-Hill publication, 1999.

156. D. Dasgupta. An Artificial Immune System as a Multiagent Decision Support System. In IEEE Int. Conf. on Systems, Man, and Cybernetics, San Diego, 1998.

157. D. Dasgupta. Artificial Neural Networks and Artificial Immune Systems: Similarities and Differences. Published in the proceedings of the IEEE International Conference on Systems, Man and Cybernetics, Orlando, October 12–15, 1997.

158. D. Dasgupta and N. Attoh-Okine. Immunity-based systems: A survey. Published in the proceedings of the IEEE International Conference on Systems, Man, and Cybernetics, pp. 363–374, Orlando, Florida, October 12–15 1997.

159. D. Dasgupta. Artificial Neural Networks vs. Artificial Immune Systems. Published in the proceedings of the Sixth International Conference on Intelligent Systems, Boston, June 11–13, 1997.

160. D. Dasgupta. A new Algorithm for Anomaly Detection in Time series Data. In International Conference on Knowledge based Computer Systems (KBCS), Bombay, India, December 16–18, 1996.

161. D. Dasgupta. Using Immunological Principles in Anomaly Detection. Published in the proceedings of the Artificial Neural Networks in Engineering (ANNIE), St. Louis, USA, November 10–13 1996.

162. D. Dasgupta and S. Forrest. Novelty Detection in Time Series Data using Ideas from Immunology. Published in the proceedings of the ISCA 5th International Conference on Intelligent Systems, Reno, Nevada, June 19–21 1996.

163. D. Dasgupta and S. Forrest. Tool Breakage Detection in Milling Operations using a Negative-Selection Algorithm. Technical Report CS95-5, Department of Computer Science, University of New Mexico, 1995.

164. Despina Davoudani, Emma Hart, Ben Paechter. Conceptual: An Immune-Inspired Approach to Speckled Computing. In the proceedings of 6th international conference on Artificial Immune systems, 26th–29th August, 2007 in Santos/SP, Brazil.

165. Pablo A. D. de Castro, Fabrício O. França, Hamilton M. Ferreira, Fernando Von Zuben. Classification and Clustering: Applying Biclustering to Text Mining: An Immune-Inspired Approach. In the proceedings of 6th international conference on Artificial Immune systems, 26th–29th August, 2007 in Santos/SP, Brazil.

166. L. N. de Castro. The Immune Response of an Artificial Immune Network (AINet). Published at the IEEE Congress on Evolutionary Computation, Canberra, Australia, December 8–12, 2003.

167. L. N. de Castro, Fundamentals of Neurocomputing, Technical Report—RT DCA 01/03, 72 p. (2003).

168. L. N. de Castro and F. J. Von Zuben. The Construction of a Boolean Competitive Neural Network Using Ideas from Immunology, (pre-print), Neurocomputing, 50C, pp. 51–85, 2003.

169. L. N. de Castro and J. Timmis, Artificial Immune Systems as a Novel Soft Computing Paradigm. In the Soft Computing Journal, Vol. 7, Issue 7, July 2003.

170. L. N. de Castro. Immune Engineering: A Personal Account, II Workshop on Computational Intelligence and Semiotics, CD ROM Proceedings (2002).

171. L. N. de Castro. Comparing immune and neural networks. Published in the proceedings of VII Brazilian Symposium on (SBRN), pp. 250–255. November 11–14, 2002.

172. L. N. de Castro and J. Timmis. Hierarchy and Convergence of Immune Networks: Basic Ideas and Preliminary Results. Published in the proceedings of 1st International Conference on Artificial Immune Systems (ICARIS), University of Kent at Canterbury, U.K., September 9th–11th, 2002.

173. L. N. de Castro and F. J. Von Zuben. Learning and Optimization Using the Clonal Selection Principle. In the Special Issue on Artificial Immune Systems of the journal IEEE Transactions on Evolutionary Computation, Vol. 6, No. 3, June 2002.

174. L. N. de Castro and J. Timmis. Artificial Immune Systems: A Novel Approach to Pattern Recognition. In L. Alonso J. Corchado and C. Fyfe, editors, *Artificial Neural Networks in Pattern Recognition*, pp. 67–84. University of Paisley, January 2002.

175. L. N. de Castro and J. Timmis. An Artificial Immune Network for Multimodal Function Optimization. Published in the proceedings of the special sessions on artificial immune systems in Congress on Evolutionary Computation, IEEE World Congress on Computational Intelligence, Honolulu, Hawaii, May 2002.

176. L. N. de Castro and F. J. Von Zuben. An Immunological Approach to Initialize Feedforward Neural Network Weights. Published at International Conference on Artificial Neural Networks and Genetic Algorithms (ICANNGA), 2001.

177. L. N. de Castro and F. J. Von Zuben. A Pruning Self-Organizing Algorithm to Select Centers of Radial Basis Function Neural Networks. Published at ICANNGA, 2001 (International Conference on Artificial Neural Networks and Genetic Algorithms).

178. L. N. de Castro and F. J. Von Zuben (2001f). The Construction of a Boolean Competitive Neural Network Using Ideas from Immunology.

179. L. N. de Castro and F. J. Von Zuben (2001g). AiNet: An Artificial Immune Network for Data Analysis. Book Chapter in Data Mining: A Heuristic Approach, Hussein A. Abbass, Ruhul A. Sarker, and Charles S. Newton (Eds.), Idea Group Publishing, USA.

180. L. N. de Castro and F. J. Von Zuben (2001h). Immune and Neural Network Models: Theoretical and Empirical Comparisons. Invited paper to the International Journal of Computational Intelligence and Applications (IJCIA).

181. L. N. de Castro and F. J. Von Zuben (2001b). An Immunological Approach to Initialize Centers of Radial Basis Function Neural Networks. (Pre-print). Published in the proceedings of CBRN'01 (Brazilian Conference on Neural Networks), pp. 79–84.

182. L. N. de Castro and F. J. Von Zuben (2000a). The Clonal Selection Algorithm with Engineering Applications. (Full version, pre-print). Published in the proceedings of Genetic and Evolutionary Computation Conference (GECCO) 2000(Workshop Proceedings), pp. 36–37.

183. L. N. de Castro and F. J. Von Zuben (2000b). An Evolutionary Immune Network for Data Clustering. (Full version, pre-print). Published in the Proceedings of the IEEE SBRN'00 (Brazilian Symposium on Artificial Neural Networks), pp. 84–89.

184. L. N. de Castro and F. J. Von Zuben (2001b). Automatic Determination of Radial Basis Function: An Immunity-Based Approach. Published in the International Journal of Neural Systems (IJNS), Special Issue on Non-Gradient Learning Techniques.

185. L. N. de Castro and F. J. Von Zuben. The Clonal Selection Algorithm with Engineering Applications. Published in the proceedings of Genetic and Evolutionary Computation Conference (GECCO), Las Vegas, Nevada, USA, July 8, 2000.

186. L. N. de Castro and F. J. Von Zuben (2000). Artificial Immune Systems: Part II – A Survey of Applications. Technical Report – RT DCA 02/00, FEEC/UNICAMP, Brazil, 64 p.

187. L. N. de Castro and F. J. Von Zuben (1999). Artificial Immune Systems: Part I – Basic Theory and Applications. Technical Report – RT DCA 01/99, FEEC/UNICAMP, Brazil, 95 p.

188. R. Deaton, M. Garzon, J. A. Rose, R. C. Murphy, S. E. Stevens, Jr and D. R. Franceschetti. DNA Based Artificial Immune System for Self-Nonself Discrimination. Published in the proceedings of the IEEE International Conference on Systems, Man, and Cybernetics, Orlando, Florida, October 13, 1997.

189. R. J. DeBoer, P. Hogeweg and A. S. Perelson. Growth and recruitment in the immune network. In A. S. Perelson and C. Weisbuch, editors, Theoretical and Experimental Insights into Immunology, pages 223–247, Springer-Verlag, Berlin, 1992.

190. R. J. DeBoer, L. A. Segel and A. S. Perelson. Pattern formation in one and two dimensional shape space models of the immune system. J. Theoret. Biol., 155:295–333, 1992.

191. R. J. DeBoer and A. S. Perelson. Size and connectivity as emergent properties of a developing immune network. J. Theoretical Biology, 149:381–424, 1991.

192. L. Honório de Mello, Armando M. Leite da Silva, Daniele A. Barbosa. Search and Optimization: A Gradient-Based Artificial Immune System Applied to Optimal Power Flow Problems. In the proceedings of 6th international conference on Artificial Immune systems, 26th–29th August, 2007 in Santos/SP, Brazil.

193. J. H. B. De Monvel and O. C. Martin. Memory capacity in large idiotypic networks. Bull. Math. Biol. 57, 109, 1995.

194. G. Danezis, G. Diaz, S. Faust, E. K¨asper, C. Troncoso C., and B. Preneel, "Efficient negative databases from cryptographic hash functions," in Information Security Conference, Springer LNCS, Ed., 2007, vol. 4779, pp. 423–436.

195. J. S. de Sousa, L. de C. T. Gomes, G. B. Bezerra, L. N. de Castro & F. J. Von Zuben (2004), An Immune-Evolutionary Algorithm for Multiple Rearrangements of Gene Expression Data, Vol. 5, pp. 157–179. Link [http://www.dca.fee.unicamp.br/~lnunes/publicat.htm]

196. V. Detours, B. Sulzer and A. S. Perelson. Size and connectivity of the idiotypic network are independent of the discreteness of the affinity distribution. J. Theoret. Biol., 183:409–416, 1996.

197. V. Detours, H. Bersini, J. Stewart and F. Varela, Development of an Idiotypic Network in Shape Space. Journal of Theoretical Biology, 170, 1994.

198. P. D'haeseleer, S. Forrest and P. Helman. A distributed approach to anomaly detection. Submitted to ACM Transactions on Information System Security, 1997.

199. P. D'haeseleer, S. Forrest and P. Helman. An immunological approach to change detection: algorithms, analysis, and implications. Published in the proceedings of the 1996 IEEE Symposium on Computer Security and Privacy, IEEE Computer Society Press, Los Alamitos, CA, pp. 110–119, 1996.

200. P. D'haeseleer. An immunological approach to change detection: Theoretical Results. In 9th IEEE Computer Security Foundations Workshop, 1996.

201. Werner Dilger. Structural Properties of Shape-Spaces. Published in the proceedings of the 5th International Conference on Artificial Immune Systems (ICARIS), Portugal, 4–6 September, 2006.

202. W. Dilger. Decentralized Autonomous Organization of the Intelligent Home According to the Principles of the Immunity System. Published in the proceedings of the IEEE International Conference on Systems, Man, and Cybernetics, Orlando, Florida, October 13, 1997.

203. W. Dilger and S. Strangfeld. Properties of the Bersini Experiment on Self-Assertion. Published in the proceedings of the Genetic and Evolutionary Computation Conference (GECCO), July 8–12, 2006.

204. Y. Ding and L. Ren. Fuzzy, Self-tuning immune feedback controller for tissue hyperthermia. IEEE International Conference on Fuzzy Systems, San Antonio. 1:534–538. 2000.

205. LIU Di and ZHU Xuefeng. Application of Immunological Memory to the Color Classification of Tiles. Published in the proceedings of the IEEE Congress on Evolutionary Computation, Canberra, Australia, December 8–12, 2003.

206. Jorge Luís M. do Amaral, José F. M. Amaral, Ricardo Tanscheit. Applications and Negative Selection: Real-Valued Negative Selection Algorithm with a Quasi-Monte Carlo Genetic Detector Generation. In the proceedings of 6th international conference on Artificial Immune systems, 26th–29th August, 2007 in Santos/SP, Brazil.

207. Shih Dong-Her, Chiang Hsiu-Sen, Chan Chun-Yuan and Binshan Lin. Internet security: malicious e-mails detection and protection. Industrial Management & Data Systems Volume 104 · Number 7 · 2004 · pp. 613–623.

208. F. Dongmei, Z Deling, Chen Ying. Design and Simulation of a Biological Immune Controller Based on Improved Varela Immune Network Model. Published in the proceedings of ICARIS, 4th International Conference on Artificial Immune Systems, Banff, Canada, 2005.

209. Gerry Dozier, Douglas Brown, Krystal Cain, John Hurley. Vulnerability Analysis of Immunity-Based Intrusion Detection Systems Using Evolutionary Hackers. Published in the proceedings of International Conference on Genetic and Evolutionary Computation (GECCO). Seattle, Washington USA, June 26–30, 2004.

210. Gerry V. Dozier, Douglas Brown, John Hurley and Krystal Cain. Swarm Intelligence: Vulnerability Analysis of AIS-Based Intrusion Detection Systems via Genetic and Particle Swarm Red Teams. Published in the proceedings of the Congress on Evolutionary Computation (CEC). Portland, Oregon USA, June 20–23, 2004.

211. Hai-Feng Du, Li-Cheng Jiao, Sun-An Wang. Clonal operator and antibody clone algorithms. Published in the proceedings of Machine Learning and Cybernetic International Conference. pp. 506–510, Vol. 1. November 4–5, 2002.

E

212. M. Ebner, Hans-Georg Breunig and J. Albert. On the Use of Negative Selection in an Artificial Immune System (MPP). Published in the proceedings of the International Conference Genetic and Evolutionary Computation (GECCO), New York, July 9–13, 2002.

213. K. N. Edge, G. B. Lamont, R. A. Raines. A Retrovirus Inspired Algorithm for Virus Detection & Optimization. Published in the proceedings of the Genetic and Evolutionary Computation Conference (GECCO), July 8–12, 2006.

214. S. Endoh, N. Tom and K. Yamada. Immune algorithm for n-tsp. Pages 3844–3849, Published in the proceedings of IEEE International Conference on Systems and Man and Cybernetics (SMC), San Diego, USA:IEEE, 1998.

215. F. Esponda, E. D. Trias, E. S. Ackley, and S. Forrest, "A relational algebra for negative databases," Technical report TR-CS-2007-18, University of New Mexico, 2007.

216. F. Esponda, E. S. Ackley, P. Helman, H. Jia, and Stephanie Forrest. Protecting data privacy through hard-to-reverse negative databases, International Journal of Information Security, Vol. 6, no. 6, pp. 403–416, October 2007.

217. F. Esponda, "Negative Surveys," ArXiv Mathematics e-prints:math/0608176, Aug. 2006.

218. F. Esponda, E. S. Ackley, S. Forrest, and P. Helman, "On-line negative databases (with experimental results)," International Journal of Unconventional Computing, Vol. 1, no. 3, pp. 201–220, 2005.

219. F. Esponda, E.S. Ackeley, S.Forrest and P. Helman. On-Line Negative Databases (Conceptual Paper). Published in the proceedings of the Third International Conference on Artificial Immune Systems (ICARIS), Catania, Italy. September 13–16, 2004.

220. F. Esponda, S. Forrest, and P. Helman, "Enhancing privacy through negative representations of data," CS Technical report, University of New Mexico, 2004.

221. Fernando Esponda and Stephanie Forrest and Paul Helman. The Crossover Closure and Partial Match Detection. Published in the proceedings of the 2nd International Conference on Artificial Immune Systems (ICARIS), pp. 249–260, 2003.

222. F. Esponda, Stephanie Forrest and Paul Helman. A Formal Framework for Positive and Negative Detection Schemes, IEEE Transactions on System, Man, and Cybernetics, 2003.

223. F. Esponda and Stephanie Forrest. Defining self: Positive and negative detection, The University of New Mexico, Albuquerque, NM, TR-CS-2002-02, 2002.

224. F. Esponda and Stephanie Forrest. Detector coverage under the r-contiguous bits matching rule, The University of New Mexico, Albuquerque, NM, TR-CS-2002-03, 2002.

F

225. J. D. Farmer. A rosetta stone for connectionism. Physica D, 42:153-187, 1990.

226. J. D. Farmer, N. H. Packard and A. S. Perelson. The immune system, adaptation, and machine learning. Physica D, 22:187–204, 1986.

227. J. Faro, Jaime Combadao, and Isabel Gordo. Did Germinal Centers evolve under differential effects of diversity vs affinity. Published in the proceedings of the 5th International Conference on Artificial Immune Systems (ICARIS), Portugal, 4–6 September, 2006.

228. J. Faro and S. Velasco. Studies on a recent class of network models of the immune system. J. Theor. Biol. 164, 271, 1993.

229. Feng-Xian Wang, Jie Zhao, Sheng Chang, Ji-Min Li, Zhen-Peng Liu. FICSEM: A learning method from one-case fitted in Complex Adaptive System. Machine Learning and Cybernetics, International Conference. pp. 1796–1800, Vol. 4. November 4–5, 2002.

230. M. A. Fishman and A. S. Perelson. Modeling T cell-antigen presenting cell interactions. J. Theoret. Biol., Vol. 160, pp. 311–342, 1993.

231. Grazziela P. Figueredo, Nelson F. F. Ebecken, Helio J. C. Barbosa. Classification and Clustering: The SUPRAIC algorithm: A Suppression Immune Based Mechanism to Find a Representative Training Set in Data Classification Tasks. In the proceedings of 6th international conference on Artificial Immune systems, 26th–29th August, 2007 in Santos/SP, Brazil.

232. Grazziela Patrocinio Figueredo, Nelson Favilla Ebecken, Helio Correa Barbosa. Suppression based immune mechanism to find a representative training set in data classification tasks. Published in the Proceedings of the 9th annual conference on Genetic and evolutionary computation (GECCO) 2007, pp. 171–171, London, England.

233. S. Forrest, J. Balthrop, M. Glickman, and D. Ackley, Computation in the Wild. K. Park and W. Willins (eds.). The Internet as a Large-Scale Complex System. Oxford University Press.

234. S. Forrest and S. A. Hofmeyr. Engineering an immune system. Graft, Vol.4: 5, pp. 5–9, 2001.

235. S. Forrest and S. A. Hofmeyr. Immunology as information processing. In Design Principles for the Immune System and Other Distributed Autonomous Systems, edited by L. A. Segel and I. Cohen. Santa Fe Institute Studies in the Sciences of Complexity. New York: Oxford University Press, 2000.

236. S. Forrest and S. A. Hofmeyr. John Holland's Invisible Hand: An Artificial Immune System, Presented at the Festschirift. 1999.

237. S. Forrest, A. Somayaji and D. H. Ackley. Building diverse computer systems. Published in the proceedings of the Sixth Workshop on Hot Topics in Operating Systems, IEEE Computer Society Press, Los Alamitos, CA, pp. 67–72, 1997.

238. S. Forrest, S. Hofmeyr and A. Somayaji. Computer Immunology. In Communications of the ACMVol. 40, No. 10, pp. 88–96, 1997.

239. S. Forrest, A. Somayaji and D. Ackley. Building Diverse Computer Systems. Published in the proceedings of the Sixth Workshop on Hot Topics in Operating Systems, 1997.

240. S. Forrest, S. A. Hofmeyr, A. Somayaji and T. A. Longstaff. A sense of self for Unix processes. Published in the proceedings of 1996 IEEE Symposium on Computer Security and Privacy, 1996.

241. S. Forrest, A. S. Perelson, L. Allen and R. Cherukuri. Self-nonself discrimination in a computer. Published in the proceedings of the IEEE Symposium on Research in Security and Privacy, IEEE Computer Society Press, Los Alamitos, CA, pp. 202–212, 1994.

242. S. Forrest, B. Javornik, R. E. Smith and A. S. Perelson. Using genetic algorithms to explore pattern recognition in the immune system. In Evolutionary Computation 1:3, pp. 191–211, 1993.

243. S. Forrest and A. S. Perelson. Genetic algorithms and the immune system. In H. Schwefel and R. Maenner (Eds.) Parallel Problem Solving from Nature, Springer-Verlag, Berlin. (Lecture Notes in Computer Science), 1991.

244. S. Forrest and A. S. Perelson. Genetic algorithm and the Immune System. Published in the proceedings of the first Workshop on Parallel Problem Solving from Nature, Dortmund, Federal Republic of Germany, 1–3, October, 1990.

245. Claudio Franceschi. The Immune System as a Cognitive System: New Perspectives for Information Technology Society. Published in the proceedings of Second International Conference on Artificial Immune Systems (ICARIS), Napier University, Edinburgh, U.K., September 1–3, 2003.

246. S. A. Frank. The Design of Natural and Artificial Adaptive Systems. Academic Press, New York, M. R. Rose and G. V. Lauder edition, 1996.

247. Alex A. Freitas and Jon Timmis. Revisiting the Foundations of Artificial Immune Systems: A Problem-oriented Perspective. Published in the proceedings of Second International Conference on Artificial Immune Systems (ICARIS), September 1–3, 2003, Napier University, Edinburgh, U.K.

248. F. Freschi and M. Repetto. Multiobjective Optimization by a Modified Artificial Immune System Algorithm. Published in the proceedings of ICARIS, 4th International Conference on Artificial Immune Systems, Banff, Canada, 2005.

249. T. Fukuda, K. Mori and M. Tsukiyama. Parallel Search for Multi-Modal Function Optimization with diversity and Learning of Immune algorithm. Chapter 11 in the book entitled Artificial Immune Systems and Their Applications, Publisher: Springer-Verlag, Inc., pp. 210–219, January 1999.

250. T. Fukuda, K. Mori and M. Tsukiyama. Immunity Based Management System for a Semiconductor Production Line. Chapter 23 in the book entitled Artificial Immune Systems and Their Applications, Publisher: Springer-Verlag, Inc., pp. 278–288, January 1999.

251. T. Fukuda, K. Mori and M. Tsukiyama. Immune Networks using Genetic Algorithm for Adaptive Production Scheduling. In 15th IFAC World Congress, Vol. 3, pp. 57–60, 1993.

G

252. Juan Carlos Galeano, Angélica Veloza-Suan and Fabio A González. A Comparative Analysis of Artificial Immune Network Models. Published in the proceedings of the Genetic and Evolutionary Computation Conference (GECCO), Washington, D.C., June 25–29, 2005.

253. Wei Gao. Fast Immunized Evolutionary Programming. [Poster] Published in the proceedings of Congress on Evolutionary Computation (CEC). Portland, Oregon USA, June 19–23, 2004

254. N. Ganguly and A. Deutsch. Developing Efficient Search Algorithms for P2P Networks Using Proliferation and Mutation. Published in the proceedings of the Third International Conference on Artificial Immune Systems (ICARIS), Catania, Italy. September 13–16, 2004.

255. Utpal Garain, Mangal P. Chakraborty and Dipankar Dasgupta Recognition of Handwritten Indic Script using Clonal Selection Algorithm. Published in the proceedings of the 5th International Conference on Artificial Immune Systems (ICARIS), Portugal, 4–6 September, 2006.

256. S. M. Garrett. A Survey of Artificial Immune Systems: Are They Useful? Evolutionary Computation, 2005.

257. Simon Garrett. Parameter-Free, Adaptive Clonal Selection. Published in the proceedings of the Congress on Evolutionary Computation (CEC). Portland, Oregon USA, June 20–23, 2004.

258. Simon M. Garrett. A Paratope is Not an Epitope: Implications for Immune Network Models and Clonal Selection. Published in the proceedings of Second International Conference on Artificial Immune Systems (ICARIS), September 1–3, 2003, Napier University, Edinburgh, U.K.

259. Gaspar & Hirsbrunner. From Optimization to Learning in Learning in Changing Environments: The Pittsburgh Immune Classifier System. Published in the proceedings of 1st International Conference on Artificial Immune Systems (ICARIS), University of Kent at Canterbury, U.K., September 9th–11th, 2002.

260. A. Gaspar and P. Collard. Two Models of Immunization for time dependent Optimization. Published in the proceedings of IEEE International Conference on Systems, Man and Cybernetics (SMC), Nashville, October 8–11, 2000.

261. A. Gaspar and P. Collard. Immune Approaches to experience acquisition in Time Dependent Optimization. Published in the proceedings of the Genetic and Evolutionary Computation Conference (GECCO), Las Vegas, Nevada, USA, July 8, 2000.

262. A. Gaspar and P. Collard. From Gas to Artificial Immune Systems: Improving adaptation in time dependent optimization. Published in the proceedings of the Congress on Evolutionary Computation, pp. 1859–1866. 1999.

263. Ge Hong, Mao Zong-Yuan. Immune algorithm. Published in the proceedings of the 4th World Congress on Intelligent Control and Automation, Vol. 3, pp. 1784–1788. June 10–14, 2002.

264. C. J. Gibert and T. W. Routen. Associative memory in an immune-based system. Published in the proceedings of the 12th National Conference on Artificial Intelligence (AAAI), pp. 852–857, Seattle, July 31–August 4, 1994.

265. M. Gilfix. An integrated Software Immune System: A Framework for Automated Network Management, System health, and Security. 24th Conference in Local Computer Networks. Lowell, Massachusetts. October 17–20, 1999.

266. J. Gomez, F. Gonzalez and D. Dasgupta, An Immuno-Fuzzy Approach to Anomaly Detection. Published in the proceedings of the IEEE International Conference on Fuzzy Systems (FUZZIEEE) May 25–28, 2003.

267. Richard A. Goncalves, Carolina P. de Almeida, Myriam R. Delgado, Elizabeth F. Goldbarg, Marco Cesar Goldbarg. General Applications: A Cultural Immune System for Economic Load Dispatch with Non-Smooth Cost Functions. In the proceedings of 6th international conference on Artificial Immune systems, 26th–29th August, 2007 in Santos/SP, Brazil.

268. Larisa Goncharova, Yuri Melnikov and Alexander Tarakanov. Biomolecular Immunocomputing. Published in the proceedings of Second International Conference on Artificial Immune Systems (ICARIS), Napier University, Edinburgh, U.K., September 1–3, 2003.

269. Maoguo Gong, Lining Zhang, Licheng Jiao and Shuiping Gou. Solving Multiobjective Clustering Using an Immune-Inspired Algorithm. In the proceedings of Congress on Evolutionary Computation (CEC) 2007, 25–28 September, Singapore.

270. Gong M, Jiao L, Liu F, Du H. The Quaternion Model of Artificial Immune Response. Published in the proceedings of ICARIS, 4th International Conference on Artificial Immune Systems, Banff, Canada, 2005.

271. Maoguo Gong, Ling Wang, Licheng Jiao, Haifeng Du. An artificial immune system algorithm for CDMA multiuser detection over multi-path channels. Published in the proceedings of the 2005 conference on Genetic and evolutionary computation GECCO, June 2005 Publisher: ACM Press.

272. Luis Gonzalez and James Cannady. A Self-adaptive Negative Selection approach for Anomaly Detection. Published in the proceedings of the Congress on Evolutionary Computation (CEC). Portland, Oregon USA, June 20–23, 2004.

273. F. González, J. Galeano, A. Veloza and A. Rojas. Neuro-Immune Model for Discriminating and Visualizing Anomalies. *Natural Computing Journal*, 5:3, pp. 285–304, Springer-Verlag, Sepember 2006.

274. F. González and D. Dagupta. Anomaly detection using real-valued negative selection. *Genetic Programming and Evolvable Machines*, 4(4), pages 383–403, Kluwer Acad. Publ., December 2003.

275. Fabio Gonzalez, Dipankar Dasgupta and Luis Fernado Nino. A Randomized Real-Valued Negative Selection Algorithm. Published in the proceedings of Second International Conference on Artificial Immune Systems (ICARIS), Napier University, Edinburgh, U.K., September 1–3, 2003.

276. F. Gonzalez, D. Dasgupta, and J. Gomez. 'The effect of binary matching rules in negative selection'. Published in the proceedings of the Genetic and Evolutionary Computation Conference (GECCO), Chicago, July 12–16, 2003. LNCS 2723, p. 195 ff.

277. Gonzalez & Dasgupta. Neuro-Immune and Self-Organizing Map Approaches to Anomaly Detection: A Comparison. Published in the proceedings of the 1st International Conference on Artificial Immune Systems (ICARIS), University of Kent at Canterbury, U.K., September 9th–11th, 2002.

278. F. Gonzalez and D. Dasgupta. An Immunogenetic Technique to Detect Anomalies in Network Traffic (RWA). Published in the proceedings of the International Conference Genetic and Evolutionary Computation (GECCO), New York, July 9–13, 2002.

279. F. Gonzalez, D. Dasgupta and R. Kozma. Combining Negative Selection and Classification Techniques for Anomaly Detection. Published in the proceedings of the special sessions on artificial immune systems in Congress on Evolutionary Computation, IEEE World Congress on Computational Intelligence, Honolulu, Hawaii, May 2002.

280. Fabio A González, Juan Carlos Galeano, Diego Alexander Rojas and Angélica Veloza-Suan. Discriminating and Visualizing Anomalies Using Negative Selection and Self-Organizing Maps. Published in the proceedings of the Genetic and Evolutionary Computation Conference (GECCO), Washington, D.C, June 25–29, 2005.

281. A. J. Graaff and A. P. Engelbrecht. A Local Network Neighbourhood Artificial Immune System for Data Clustering. In the proceedings of Congress on Evolutionary Computation (CEC) 2007, 25–28 September, Singapore.

282. A. J. Graaff, A. P. Engelbrecht. Using a threshold function to determine the status of lymphocytes in the artificial immune system. Published in the proceedings of the Annual research conference of the South African institute of computer scientists and information technologists on Enablement through technology (SAICSIT), Publisher: South African Institute for Computer Scientists and Information Technologists, September 2003.

283. Julie Greensmith, Uwe Aickelin and Jamie Twycross. Articulation and Clarification of the Dentritic Cell Algorithm. Published in the proceedings of the 5th International Conference on Artificial Immune Systems (ICARIS), Portugal, 4–6 September, 2006.

284. J. Greensmith, U. Aickelin and S. Cayzer. Introducing Dendritic Cells as a Novel Immune-Inspired Algorithm for Anomaly Detection. Published in the proceedings of ICARIS, 4th International Conference on Artificial Immune Systems, Banff, Canada, 2005.

285. Julie Greensmith and Steve Cayzer. An Artificial Immune System Approach To Semantic Document Classification. Published in the proceedings of Second International Conference on Artificial Immune Systems (ICARIS), September 1–3, 2003, Napier University, Edinburgh, U.K.

286. A. Grillo, A. Caetano and A. Rosa. Agent based Artificial Immune System. Published in the proceedings of Genetic and Evolutionary Computation Conference GECCO, (Late Breaking Papers.) pp. 145–151, San Francisco, California, July 9–11, 2001.

287. A. Grillo, A. Caetano. Immune System Simulation through a Complex Adaptive System Model. Published in the proceedings of the Third workshop on Genetic Algorithm and Artificial Life- GAAL'99, 1999.

288. J. Gu, D. Lee, S. Park and K. Sim. An Immunity-based Security Layer Model. Workshop on Artificial Immune System at Genetic and Evolutionary Computation Conference GECCO, Las Vegas, Nevada, USA, July 8, 2000.

289. Z. Guo and J. C. Tay. A Comparative Study on Modeling Strategies for Immune System Dynamics under HIV-1 Infection. Published in the proceedings of ICARIS, 4th International Conference on Artificial Immune Systems, Banff, Canada, 2005.

290. Zaiyi Guo, Hann Kwang Han and Joc Cing Tay. Sufficiency Verification of HIV-1 Pathogenesis based on Multi-Agent Simulation. Published in the proceedings of the Genetic and Evolutionary Computation Conference (GECCO), Washington, D.C., June 25–29, 2005.

291. Thiago Guzella, Tomaz Mona-Santos, Joacquim Uchoa and Walmir Caminhas. Modelling the Control of an Immune Response through Cytokine Signalling. Published in the proceedings of the 5th International Conference on Artificial Immune Systems (ICARIS), Portugal, 4–6 September, 2006.

292. Thiago S. Guzella, Tomaz A. Mota-Santos, Walmir M. Caminhas. Applications and Anomaly Detection: A Novel Immune Inspired Approach to Fault Detection. In the proceedings of 6th international conference on Artificial Immune systems, 26th–29th August, 2007 in Santos/SP, Brazil.

293. Thiago S. Guzella, Tomaz A. Mota-Santos, Walmir M. Caminhas. Applications and Anomaly Detection: Towards a Novel Immune Inspired Approach to Temporal Anomaly Detection. In the proceedings of 6th international conference on Artificial Immune systems, 26th–29th August, 2007 in Santos/SP, Brazil.

294. Thiago S. Guzella, Tomaz A. Mota-Santos, Walmir M. Caminhas. Conceptual: Regulatory T Cells: Inspiration for Artificial Immune Systems. In the proceedings of 6th international conference on Artificial Immune systems, 26th–29th August, 2007 in Santos/SP, Brazil.

H

295. Charles R. Haag, Gary B. Lamont, Paul D. Williams, Gilbert L. Peterson. Applications and Anomaly Detection: An Artificial Immune System-Inspired Multiobjective Evolutionary Algorithm with Application to the Detection of Distributed Computer Network Intrusions. In the proceedings of 6th international conference on Artificial Immune systems, 26th–29th August, 2007 in Santos/SP, Brazil.

296. Kiryong Ha, Inho Park, Jeonwoo Lee, Doheon Lee. Applications and Anomaly Detection: Automated Blog Design System with a Population-Based Artificial Immune Algorithm. In the proceedings of 6th international conference on Artificial Immune systems, 26th–29th August, 2007 in Santos/SP, Brazil.

297. P. Hajela and J. S. Yoo. Immune Network Modeling in Design Optimization. A chapter in the book, "New Ideas in Optimization" pp. 203–215. McGraw–Hill, 1999.

298. P. Hajela and J. Lee. Constrained Genetic Search Via Schema Adaptation: An Immune Network Solution. Structural Optimization, Vol. 12, No. 1, pp. 11–15, 1996.

299. P. Hajela, J. Yoo and J. Lee. GA Based Simulation of Immune Networks - Applications in Structural Optimization. Journal of Engineering Optimization, 1997.

300. Ramin Halavati, Saeed Bagheri Shouraki, Mojdeh Jalali Heravi, Bahareh Jafari Jashmi. An artificial immune system with partially specified antibodies. Published in the Proceedings of the 9th annual conference on Genetic and evolutionary computation (GECCO '07), pp. 57–62, London, England.

301. Janna Hamaker and Lois Boggess. Non-Euclidean Distance Measures in AIRS, an Artificial Immune Classification System. Published in the proceedings of the Congress on Evolutionary Computation (CEC). Portland, Oregon USA, June 20–23, 2004.

302. Xiaoshu Hang and Honghua Dai. Applying both Positive and Negative Selection to Supervised Learning for Anomaly Detection. Published in the proceedings of the Genetic and Evolutionary Computation Conference (GECCO), Washington, D.C., June 25–29 2005.

303. Xiaoshu Hang, Honghua Dai. Combining Computational Immunology and Coevolutionary GA for Anomaly Detection. Published in the proceedings of International Conference on Genetic and Evolutionary Computation (GECCO) Seattle, Washington USA, June 26–30, 2004.

304. Xiaoshu Hang, Honghua Dai. Constructing Detectors in Schema Complementary Space for Anomaly Detection. Published in the proceedings of International Conference on Genetic and Evolutionary Computation (GECCO). Seattle, Washington USA, June 26–30, 2004.

305. Emma Hart, Francisco Santos, Hugues Bersini. Modeling: Topological Constraints in the Evolution of Idiotypic Networks. In the proceedings of 6th international conference on Artificial Immune systems, 26th–29th August, 2007 in Santos/SP, Brazil.

306. Emma Hart - Analysis of a Growth Model for Idiotypic Networks. Published in the proceedings of the 5th International Conference on Artificial Immune Systems (ICARIS), Portugal, 4–6 September, 2006.

307. Emma Hart, Hugues Bersini and Francisco Santos—Tolerance vs. Intolerance: How Affinity Defines Topology in an Idiotypic Network. Published in the proceedings of the 5th International Conference on Artificial Immune Systems (ICARIS), Portugal, 4–6 September, 2006.

308. E. Hart and J. Timmis. Application Areas of AIS: The Past, The Present and The Future. Published in the proceedings of ICARIS, 4th International Conference on Artificial Immune Systems, Banff, Canada, 2005.

309. E. Hart. Not All Balls Are Round: An Investigation of Alternative Recognition-Region Shapes. Published in the proceedings of International Conference on Artificial Immune Systems, 4th International Conference on Artificial Immune Systems, Banff, Canada, 2005.

310. E. Hart. Exploiting the Analogy between the Immune System and Sparse Distributed Memory. In special issues of Genetic Programming and Evolvable Machine, 2005.

311. E. Hart and P. Ross. Studies on the implications of Shape-Space Models for Idiotypic Networks. Published in the proceedings of the Third International Conference on Artificial Immune Systems (ICARIS), Catania, Italy. September 13–16, 2004.

312. E. Hart, Peter Ross, Andrew Webb, Alistair Lawson. A Role for Immunology in "Next Generation" Robot Controllers. Published in the proceedings of Second International Conference on Artificial Immune Systems (ICARIS), Napier University, Edinburgh, U.K., September 1–3, 2003.

313. Emma Hart and Peter Ross. Improving SOSDM: Inspirations from the Danger Theory Published in the proceedings of International Conference on Artificial Immune Systems (ICARIS), Napier University, Edinburgh, U.K., September 1–3, 2003.

314. E. Hart & P. Ross. Exploiting the analogy between immunology and sparse distributed memories: a system for clustering non-stationary data. Published in the proceedings of 1st International Conference on Artificial Immune Systems (ICARIS), University of Kent at Canterbury, U.K., September 9–11, 2002.

315. E. Hart and P. Ross. An Immune System Approach to Scheduling in Changing Environments. International Conference on Genetic and Evolutionary Computation, 1999.

316. E. Hart and P. Ross. The Evolution and Analysis of a potential Antibody Library for use in Job- Shop Scheduling. A chapter in the book "New Ideas in Optimization", pp. 185–202. McGraw-Hill, 1999.

317. E. Hart, P. Ross and J. Nelson. Producing Robust Schedules via an Artificial Immune System. IEEE International Conference on Evolutionary Computing, 1998.

318. P. K. Harmer, P. D. Williams, G. H. Gunsch and G. B. Lamont. An Artificial Immune System Architecture for Computer Security Applications. In the Special Issue on Artificial Immune Systems of the journal IEEE Transactions on Evolutionary Computation, Vol. 6, No. 3, June 2002.

319. P. K. Harmer. Distributed agent architecture for a computer virus immune system. M. S thesis. Air Force Institute of Technology. WPAFB. OH. March 2000. AFIT/GCE/ENG/00M-02.

320. P. K. Harmer and G. B. Lamont. Agent Based Architecture for a Computer Virus Immune System. Published in the proceedings of Genetic and Evolutionary Computation Conference GECCO, Las Vegas, Nevada, USA, July 8, 2000.

321. Y. Hasegawa and H. Iba. Multimodal Search with Immune based Genetic Programming (Conceptual Paper). Published in the proceedings of the Third International Conference on Artificial Immune Systems (ICARIS), Catania, Italy. September 13–16, 2004.

322. Xingshi He and Lin Han. A Novel Binary Differential Evolution Algorithm Based on Artificial Immune System. In the proceedings of Congress on Evolutionary Computation (CEC) 2007, 25–28 September, Singapore.

323. S. Hedberg. Combating computer viruses: IBM's new Computer Immune System. Parallel & Distributed Technology: Systems & Applications, IEEE [see also IEEE Concurrency], pp. 9–11, Vol. 4, No. 2, summer 1996.

324. P. Helman and S. Forrest. An Efficient Algorithm for Generating Random Antibody Strings. Technical Report 94–07, University of New Mexico, Albuquerque, NM, 1994.

325. R. Hightower, S. Forrest and A. S. Perelson. The Baldwin effect in the immune system: learning by somatic hypermutation. In R. K. Belew and M. Mitchell (Eds.) Adaptive Individuals in Evolving Populations, Addison-Wesley, Reading, MA, pp. 159–167, 1996.

326. R. Hightower, S. Forrest and A. S. Perelson. The evolution of emergent organization in immune system gene libraries. In L. J. Eshelman (Ed.) Published in the proceedings of the Sixth International Conference on Genetic Algorithms, Morgan Kaufmann, San Francisco, CA, pp. 344–350, 1995.

327. R. Hightower, S. Forrest, and A. S. Perelson. The evolution of secondary organization in immune system gene libraries. Published in the proceedings of the Second European Conference on Artificial Life, 1994.

328. R. Hightower, S. Forrest and A. S. Perelson. The evolution of cooperation in immune system gene libraries. Technical Report CS-92-20, University of New Mexico, Albuquerque, NM, 1992.

329. H. Hirayama and Y. Fukuyama. Analysis of dynamical transition of immune reactions of idiotype network. Presented at ICMAS Workshop on Immunity-Based Systems held on December 10, 1996.

330. H. Hirayarma and Y. Fukuyarna. A Priority in Immune System—A Hypothetical Theoretical Study. Presented at ICMAS Workshop on Immunity-Based Systems held on December 10, 1996.

331. S. A. Hofmeyr. An Interpretative Introduction to the Immune System. In Design Principles for the Immune System and other Distributed Autonomous Systems. I. Cohen and L. Segel (eds.). Oxford University Press, 2000.

332. S. Hofmeyr and S. Forrest. Intrusion Detection: Architecture for an Artificial Immune System Evolutionary Computation Journal, 2000.

333. S. A. Hofmeyr and S. Forrest. Immunity by Design: An Artificial Immune System. Published in the proceedings of the Genetic and Evolutionary Computation Conference (GECCO), San Francisco, CA, pp. 1289–1296, 1999.

334. S. A. Hofmeyr and S. Forrest. Immunity by Design: An Artificial Immune System. Published in the proceedings of Genetic and Evolutionary Computation Conference (GECCO), 1999.

335. S. A. Hofmeyr, S. Forrest, and P. D'haeseleer. An Immunological Approach to Distributed Network Intrusion Detection. Paper presented at RAID'98—First International Workshop on the Recent Advances in Intrusion Detection Louvain-la-Neuve, Belgium September 1998.

336. S. A. Hofmeyr, A. Somayaji and S. Forrest. Intrusion Detection using Sequences of System Calls. Journal of Computer Security Vol. 6, 1998. pp. 151–180.

337. S. Hofmeyr, S. Forrest and A. Sornayaji. Lightweight intrusion detection for networked operating systems. Journal of Computer Security. July 1997.

338. G. W. Hoffmann. A neural network model based on the analogy with the immune system. Journal of Theoretical Biology, 122:33–67, 1986.

339. A. Hone and J. Kelsey. Optima, extrema and artificial immune systems (Conceptual paper). Published in the proceedings of the Third International Conference on Artificial Immune Systems (ICARIS), Catania, Italy. September 13–16, 2004.

340. W. S. Hortos. Artificial immune system for securing mobile ad hoc networks against intrusion attacks. SPIE, Orlando, 2003.

341. Haiyu Hou, Gerry Dozier. Immunity-based intrusion detection system design, vulnerability analysis, and GENERTIA's genetic arms race. Published in the proceedings of the ACM symposium on Applied Computing, Santa Fe, New Mexico pp. 952–956, March 13–17, 2005.

342. Haiyu Hou, Jun Zhu, G. Dozier. Artificial immunity using constraint-based detectors. Published in the proceedings of the 5th Biannual World Automation Congress, 2002, pp. 239–244, Vol. 13, June 9–13, 2002.

343. Min Huang, Wei Tong, Qing Wang, Xin Xu and Xingwei Wang. Immune Algorithm Based Routing Optimization in Fourth-party Logistics. Published in the proceedings of IEEE World Congress on Computational Intelligence in Congress on Evolutionary Computation, Vancouver, Canada, July 16–21, 2006.

344. Chien-Feng Huang. Using an Immune System Model to Explore Mate Selection in Genetic Algorithms. Published in the proceedings of the Genetic and Evolutionary Computation Conference (GECCO), Chicago, IL, USA, July 12–16, 2003.

345. S. Huang. Immune-based optimization method to capacitor placement in a radial distribution system. IEEE Transactions on Power Delivery, Piscataway. 15(2)(2000): pp. 744–749

346. J. Hunt, J. Timmis, D. Cooke, M. Neal and C. King, Jisys: The development of an Artificial Immune System for real world applications. A chapter in the book Artificial Immune Systems and Their Applications, D. Dasgupta Ed., pp. 157–186. Pub. Springer-Verlag, 1999.ISBN 3-540-64390-7.

347. J. Hunt and J. Timmis. Evolving and Visualizing a Case Database using an Immune Network. In European Conference on Artificial Intelligence (ECAI 98), 1998.

348. J. E. Hunt, C. M. King and D. E. Cooke. Immunizing against fraud. Published in the proceedings of Knowledge Discovery and Data Mining, IEEE Colloquiurn, October 1996.

349. J. E. Hunt and D. E. Cooke. Learning Using An Artificial Immune System. In Journal of Network and Computer Applications: Special Issue on Intelligent Systems: Design and Application, Vol. 19, pp. 189–212, 1996.

350. J. E. Hunt and A. Fellows. Introducing an Immune Response into a CBR system for Data Mining. In BCS ESG'96 Conference and published as Research and Development in Expert Systems XIII, 1996.

351. J. Hunt and D. Cooke. The ISYS Project: An introduction. Tech Report. IP-REP-002, Univ. of Wales, Aberystwyth, Penglias, Aberystwyth, Dyfed, U.K., March 1996.

352. J. E. Hunt, D. E. Cooke and H. Holstein. Case memory and retrieval based on the Immune System. Published in the proceedings of the First International Conference on Case Based Reasoning, Published as Case-Based Reasoning Research and Development, Ed. Manuela Weloso and Agnar Aamodt, Lecture Notes in Artificial Intelligence 1010, pp. 205–216, October 1995.

353. J. E. Hunt and D. E. Cooke, An Adaptive and distributed Learning System based on the Immune System. Published in the proceedings of the IEEE International Conference on Systems Man and Cybernetics, pp. 2494–2499, 1995.

I

354. S. Ichikawa, A. Ishiguro, Y. Watanabe and Y. Uchikawa. Moderationism in the Immune System: Gait Acquisition of a Legged Robot Using the Metadynamics Function. In IEEE Int. Conf. on Systems, Man, and Cybernetics, San Diego, 1998.

355. Park Inho, Dokyun Na, Kwang H. Lee, Doheon Lee. Fuzzy Continuous Petri Net-Based Approach for Modeling Helper T Cell Differentiation. Published in the proceedings of ICARIS, 4th International Conference on Artificial Immune Systems, Banff, Canada, 2005.

356. Hajime Inoue and Stephanie Forrest, Anomaly Intrusion Detection in Dynamic Execution Environments, Published in the proceedings of the New Security Paradigms Workshop, 2003.

357. A. Iqbal and Maarof M. A. Polymorphism and Danger Susceptibility of System Call DASTONs. Published in the proceedings of ICARIS, 4th International Conference on Artificial Immune Systems, Banff, Canada, 2005.

358. A. Iqbal and M.A. Maarof. Towards Danger Theory based Artificial APC Model Metaphor for Danger Susceptible Data Condons (Conceptual Paper). Published in the proceedings of the Third International Conference on Artificial Immune Systems (ICARIS), Catania, Italy. September 13–16, 2004.

359. Y. Ishida. Immunity based systems: a specification and applications. Medical Imaging technology. 18(5): pp. 703–708, 2000.

360. Y. Ishida. Active diagnosis by self-organization: An approach by the immune network metaphor. Published in the proceedings of the International Joint Conference on Artificial Intelligence. Nagoya, Japan. IEEE, pp. 1084–1089, 1997.

361. Y. Ishida. The Immune System as a Self-Identification Process: a Survey and a Proposal. Published in the proceedings of ICMAS International Workshop on Immunity-Based Systems (IMBS96), Kyoto, December 10–13, pp. 2–12, 1996.

362. Y. Ishida. Distributed and autonomous sensing based on immune network, pp. 214–217 of: Published in the proceedings of Artificial Life and Robotics. Beppu. AAAI Press, 1996.

363. Y. Ishida and N. Adachi. Active Noise Control by an Immune Algorithm: Active Noise Control by an Immune Algorithm: Adaptation in Immune System as an Evolution. Published in the proceedings of ICEC 96, pp. 150–153.

364. Y. Ishida and N. Adachi. An Immune Algorithm for Multiagent: Application to Adaptive Noise Neutralization. Published in the proceedings of IROS 96. pp. 1739–1746, 1996.

365. Y. Ishida. Fully Distributed Diagnosis by PDP Learning Algorithm: Towards Immune Network PDP Model. Published in the proceedings of International Joint Conference on Neural Networks, pp. 777–782 San Diego, 1990.

366. A. Ishiguro, S. Ichikawa, T. Shibat and Y. Uchikawa. Modernationsim in the immune system: Gait acquisition of a legged robot using the metadymics function. Published in the proceedings of IEEE International Conference on Systems and Man and Cybernetics (SMC), Pages 3827–3832, San Diego, USA:IEEE, 1998.

367. A. Ishiguro, T. Kondo, Y. Watanabe, Y. Shirai and Y. Ichikawa. Emergent Construction of Artificial Immune Networks for Autonomous Mobile Robots. Published in the proceedings of SMC, pp. 1222–1228, 1997.

368. A. Ishiguro, Y. Watanabe, T. Kondo, Y. Shirai and H. Uchikawa. Immunoid: A Robot with a Decentralized Behavior Arbitration Mechanisms Based on the Immune System. Presented at ICMAS Workshop on Immunity-Based Systems, December 10, 1996.

369. A. Ishiguro, T. Kondo, Y. Watanabe and Y. Uchikawa. Immunol: An Immunological Approach to Decentralized Behavior Arbitration of Autonomous Mobile Robots. In Lecture Notes in Computer Science, Vol. 1141, Springer, pp. 6W675, 1996.

370. A. Ishiguro, S. Kuboshiki, S. Ichikawa and Y. Uchikawa. Gait Control of Hexapod Walking Robots Using Mutual-coupled Immune Networks. In Advanced Robotics, Vol. 10, No. 2, pp. 179–195, 1996.

371. A. Ishiguro, Y. Shirai, T. Kendo and Y. Uchikawa. Immunoid: An architecture for Behavior Arbitration Based on the Immune Networks. Published in the proceedings of IROS, pp. 1730–1738, 1996.

372. A. Ishiguro, Y. Watanabe and Y. Uchikawa. An Immunological Approach to Dynamic Behavior Control for Autonomous Mobile Robots. Published in the proceedings of IROS, Vol. 1, pp. 495–500, 1995.

373. A. Ishiguro, T. Kondo, Y. Watanabe and Y. Uchikawa. Dynamic Behavior Arbitration of Autonomous Mobile Robots Using Immune Networks. Published in the proceedings of ICEC, Vol. 2, pp. 722–727, 1995.

374. A. Ishiguro, S. Ichikawa and Y. Uchikawa. A Gait Acquisition of 6-Legged Walking Robot Using Immune Networks, Published in the proceedings of IRO.5'94, Vol. 2, pp. 1034–1041, 1994.

J

375. J. T. Jackson, G. H. Gunsch, R. L. Claypoole, Jr., G. B. Lamont, Blind Steganography Detection Using a Computational Immune System Approach: A Proposal. Digital Forensic Research Workshop, August 7–9, 2002.

376. Jacob T Jackson, Gregg H. Gunsch, Roger L. Claypoole, Jr., and Gary B. Lamont. Novel Steganography Detection Using an Artificial Immune System Approach. Published in the proceedings of the 2003 IEEE Congress on Evolutionary Computation, Canberra, Australia. December 8–12, 2003.

377. Christian Jacob, Scott Steil and Karel Bergmann. The Swarming Body: Simulating the Decentralized Defenses of Immunity. Published in the proceedings of the 5th International Conference on Artificial Immune Systems (ICARIS), Portugal, 4–6 September, 2006.

378. C. Jacob, J. Litorco and L. Lee. Immunity through Swarms: Agent-Based Simulations of the Human Immune System (Conceptual Paper). Published in the proceedings of the Third International Conference on Artificial Immune Systems (ICARIS), Catania, Italy. September 13–16, 2004.

379. James Graham and Yingbing Yu, "Computer System Security Threat Evaluation Based Upon Artificial Immunity Model and Fuzzy Logic," IEEE SMC (International Conference on Systems, Man and Cybernetics), Hawaii, USA - October 10–12, 2005.

380. Jr. C. A. Janeway, P. Travers with assistance of S. Hunt, M. Walport. Immunobiology: The Immune System in Health and Disease. Garland Pub. 1997.

381. Jr. C. A. Janeway. How the immune system recognizes invaders. Scientific American, 269(3): pp. 72–79, September 1993.

382. M. A. Janssen and D. W. Stow. An Application of Immunocomputing to the Evolution of Rules for Ecosystem Management. Published in the proceedings of the special sessions on artificial immune systems in Congress on Evolutionary Computation, IEEE World Congress on Computational Intelligence, Honolulu, May 2002.

383. M. A. Janssen. An immune system perspective on ecosystem management. Conservation Ecology 5(1): 13, 2001. [Online] URL: http://www.consecol. org/vol5/iss1/art13.

384. N. K. Jerne. The generative grammar of the immune system. The EMBO Journal, 4(4): pp. 847–852, 1985.

385. N. K. Jerne. Idiotopic Network and Other preconceived ideas. Immunological Rev., Vol. 79, pp. 5–24, 1984.

386. N. K. Jerne. Towards a network theory of the immune system. Ann. Immunol. (Inst. Pasteur), 125C: pp. 373–389, 1974.

387. N. K. Jerne. Clonal Selection in a lymphocyte network. Cellular Selection and Regulation in the Immune Response, pp. 39–48. Raven Press. 1974.

388. N. K. Jerne. The immune system. Scientific American, 229(l): pp. 52–60, 1973.

389. Zhou Ji and Dipankar Dasgupta. "Revisiting Negative Selection Algorithms" To appear in Evolutionary Computation Journal, Issue no. 15.2, Summer 2007.

390. Zhou Ji, Dipankar Dasgupta, Zhiling Yang and Hongmei Teng. Analysis of Dental Images Using Artificial Immune Systems. Published in the proceedings of IEEE World Congress on Computational Intelligence (special session on recent development in artificial immune systems) in Congress on Evolutionary Computation, Vancouver, Canada, July 16–21, 2006.

391. Zhou Ji and Dipankar Dasgupta. Applicability Issues of the Real-Valued Negative Selection Algorithms. (Received the Best Paper award) Published in the proceedings of International Conference on Genetic and Evolutionary Computation (GECCO). Seattle, Washington USA, July 8–12, 2006.

392. Zhou Ji, Dipankar Dasgupta. Real-Valued Negative Selection Algorithm with Variable-Sized Detectors. Published in the proceedings of International Conference on Genetic and Evolutionary Computation (GECCO), Seattle, Washington USA, June 26–30, 2004.

393. Zhou Ji and Dipankar Dasgupta. Augmented Negative Selection Algorithm with Variable-Coverage Detectors. Published in the proceedings of the Congress on Evolutionary Computation (CEC). Portland, Oregon, USA, June 19–23, 2004.

394. Zhou Ji and Dipankar Dasgupta. Estimating the Detector Coverage in a Negative Selection Algorithm. Published in the proceedings of the Genetic and Evolutionary Computation Conference (GECCO), Washington, D.C., June 25–29, 2005.

395. L. C. Jiao and L. Wang. A novel genetic algorithm based on immunity. IEEE Trans. Systems, Man and Cybernetics. 30(5): pp. 552–561. 2000.

396. Johnny Kelsey, Jon Timmis and Andrew Hone. Chasing Chaos. Published as a conceptual Paper for publication at the IEEE Congress on Evolutionary Computation, Canberra, Australia. December 8–12, 2003.

397. José Carlos L. Pinto and Zuben F. J. V. Fault Detection Algorithm for Telephone Systems Based on the Danger Theory. Published in the proceedings of ICARIS, 4th International Conference on Artificial Immune Systems, Banff, Canada, 2005.

398. K. R. L. Juca, J. B. M. Sorbral, A. Boukerche. Intrusion Detection Based on the Immune Human System. International Parallel and Distributed Processing Symposium: IPDPS workshops. Fort Lauderdale, Florida. April 15–19, 2002.

399. Jun-Zhong Zhao, Hou-Kuan Huang. An intrusion detection system based on data mining and immune principles. Published in the proceedings Machine Learning and Cybernetics International Conference, Vol. 1, pp. 524–528. November 4–5, 2002.

400. Cynthia Junqueira, Fabricio O. de Franca, Romis R. F. Attux, Cristiano M. Panazio and Leandro N. de Castro. Immune-inspired Dynamic Optimization for Blind Spatial Equalization in Undermodeled Channels. Published in the proceedings of IEEE World Congress on Computational Intelligence/Congress on Evolutionary Computation, Vancouver, Canada, July 16–21, 2006.

K

401. Johan Kaers, Richard Wheeler and Herman Verrelst. The Effect of Antibody Morphology on Non-Self Detection. Published in the proceedings of Second International Conference on Artificial Immune Systems (ICARIS), Napier University, Edinburgh, U.K. September 1–3, 2003.

402. Kaers, Wheeler & Verrelst. Building a Robust Distributed Artificial Immune Systems. In 1st International Conference on Artificial Immune Systems (ICARIS), University of Kent at Canterbury, U.K., September 9–11, 2002.

403. Vassilios Karakasis and Andreas Stafylopatis. Data Mining Based on Gene Expression Programming and Clonal Selection. Published in the proceedings of IEEE World Congress on Computational Intelligence (special session on recent development in artificial immune systems) in Congress on Evolutionary Computation, Vancouver, Canada, July 16–21, 2006.

404. M. Kayama, Y. Sugita, Y. Morooka and S. Fukuodka. Distributed diagnosis system combining the immune network and learning vector quantization. Published in the proceedings of IEEE 21st International Conference on Industrial Electronics and Control and Instrumentation. Orlando, USA. IEEE. Pages 1531–1536. 1995.

405. J. Kelsey and J. Timmis. Immune Inspired Somatic Contiguous Hypermutation for Function Optimisation. Published in the proceedings of the Genetic and Evolutionary Computation Conference (GECCO), Chicago, IL, USA, LNCS 2723, p. 207 ff. July 12–16, 2003.

406. C. Kennedy. Evolution of Self-Definition. Published in the proceedings of the IEEE Int. Conf. on Systems, Man, and Cybernetics, San Diego, 1998.

407. J. O. Kephart, G. B. Sorkin, M. Swimmer, S. R. White. Blueprint for a Computer Immune System. Chapter 21 in the book entitled Artificial Immune Systems and Their Applications, Publisher: Springer-Verlag, Inc., pp. 242–259, January 1999.

408. J. O. Kephart et al. Biologically inspired defenses against computer viruses, Published in the proceedings of IJCA 1 '95, 985–996, Montreal, August 19–25, 1995.

409. J. O. Kephart. A biologically inspired immune system for computers, in R. A. Brooks and P. Maes, eds., Artificial Life IV. Published in the proceedings of the 4th International Workshop on the Synthesis and Simulation of Living Systems, 130–139. MIT Press, 1994.

410. T. B. Kepler and A. S. Perelson. Modeling and optimization of populations subject to time-dependent mutation. Published in the proceedings of Natl. Acad. Sci. USA, 92:8219–8223, 1995.

411. T. B. Kepler and A. S. Perelson. Somatic Hypermutation in B-Cells: An optimal Control Treatment. Journal of Theoretical Biology, 164. pp. 37–64, 1993.

412. L. Kesheng, Z. Jun, C. Xianbin, W. Xufa. An algorithm based on immune principle adopted in controlling behavior of autonomous mobile robots. Computer Engineer and Applications. (5): pp. 30–32, 2000.

413. Jungwon Kim, Peter Bentley, Uwe Aickelin, Julie Greensmith, Gianni Tedesco, Jamie Twycross (2007): 'Immune System Approaches to Intrusion Detection - A Review', Natural Computing, 6(4), pp. 413–466, doi: 10.1007/s11047-006-9026-4, Abstract [http://www.cs.nott.ac.uk/%7Euxa/papers/07naco_ais_ids_review.html], Paper [http://www.cs.nott.ac.uk/%7Euxa/papers/07naco_ais_ids_review.pdf].

414. Jungwon Kim, Peter Bentley, Christian Wallenta, Mohamed Ahmed and Stephen Hailes. Danger is Ubiquitous: Detecting Malicious Activities in Sensor Networks using the Dentritic Cell Algorithm. Published in the proceedings of the 5th International Conference on Artificial Immune Systems (ICARIS), Portugal, 4–6 September, 2006.

415. J. Kim, W. Wilson, U. Aickelin and J. McLeod. Cooperative Automated worm Response and Detection ImmuNe ALgorithm (CARDINAL) inspired by T-cell Immunity and Tolerance. Published in the proceedings of ICARIS, 4th International Conference on Artificial Immune Systems, LNCS, Banff, Canada, 2005.

416. Jungwon Kim, Arlene Ong and Richard E Overill. Design of an Artificial Immune System as a Novel Anomaly Detector for Combating Financial Fraud in the Retail Sector. Published at the 2003 IEEE Congress on Evolutionary Computation, Canberra, Australia. December 8–12, 2003.

417. J. Kim and P. Bentley. A Model of Gene Library Evolution in the Dynamic Clonal Selection Algorithm.1st International Conference on Artificial Immune Systems (ICARIS), University of Kent at Canterbury, U.K., September 9th–11th, 2002.

418. J. Kim and P. Bentley. Immune Memory in the Dynamic Clonal Selection Algorithm. 1st International Conference on Artificial Immune Systems (ICARIS), University of Kent at Canterbury, U.K., September 9th–11th, 2002.

419. J. Kim and P. Bentley. Toward an Artificial Immune System for Network Intrusion Detection: An Investigation of Dynamic Clonal Selection. Published in the proceedings of the special sessions on artificial immune systems in Congress on Evolutionary Computation, IEEE World Congress on Computational Intelligence, Honolulu, Hawaii, May 2002.

420. J. Kim and P. Bentley. Evaluating negative Selection in an Artificial Immune System for Network Intrusion Detection. Genetic and Evolutionary Computation Conference GECCO 2001.

421. J. Kim and P. Bentley. Towards an Artificial Immune System for Network Intrusion Detection: An investigation of Clonal Selection with a negative Selection Operator. Published in the proceedings of the Congress on Evolutionary Computation. (CEC), Seoul, Korea, May 27–30, 2001.

422. J. Kim and P. Bentley. Negative Selection and Niching by an artificial immune system for network intrusion detection. Late Breaking Papers, Genetic and Evolutionary Computation Conference GECCO. Orlando, USA. Morgan-Kaufmann. 1999.

423. J. Kim and P. Bentley. The Artificial Immune Model for Network Intrusion Detection. 7th European Congress on Intelligent Techniques and Soft Computing (EUFIT). Aachen. Germany. September 13–19, 1999.

424. J. Kim and P. Bentley. The human Immune system and Network Intrusion Detection. Proceedings of 7th European Congress on Intelligent techniques—Soft Computing (EUFIT). Aachan. Germany. September 13–19, 1999.

425. Dong Hwa Kim. Tuning of a PID controller using an artificial immune network model and local fuzzy set. Published in the proceedings of International Conference IFSA World Congress and 20th NAFIPS, Vol. 5, pp. 2698–2703. July 25–28, 2001.

426. Helder Knidel, Fernando Von Zuben and Leandro Nunes de Castro. A Supervised Constructive Neuro-immune Network for Pattern Classification. Published in the proceedings of IEEE World Congress on Computational Intelligence/Congress on Evolutionary Computation, Vancouver, Canada, July 16–21, 2006.

427. Helder Knidel, Leandro N. de Castro, Fernando J. Von Zuben. RABNET: a real-valued antibody network for data clustering. Published in the proceedings of the conference on Genetic and evolutionary computation GECCO, June 2005 Publisher: ACM Press.

428. T. Knight and J. Timmis. Comparison of a Multi-Layered Artificial Immune System with a Kohonen Network. In the proceedings of Congress on Evolutionary Computation (CEC) 2007, 25–28 September, Singapore.

429. T. Knight and J. Timmis. Assessing the performance of the resource limited artificial immune system AINE. Technical Report 3–01, Canterbury, Kent. CT2 7NF, May 2001.

430. A. Ko, H.Y.K. Lau, T.L. Lau. General Suppression Control Framework: Application in Self-balancing Robots. Published in the proceedings of ICARIS, 4th International Conference on Artificial Immune Systems, Banff, Canada, 2005.

431. A. Ko, H. Y. K. Lau and T.L. Lau. An Immuno Control Framework for Decentralized Mechatronic Control. Published in the proceedings of the Third International Conference on Artificial Immune Systems (ICARIS), Catania, Italy. September 13–16, 2004.

432. H. Koko, M. Skok, Skrlec, D. Artificial immune systems in solving routing problems. EUROCON. Computer as a Tool. The IEEE Region 8, pp. 62–66, Vol. 1, September 22–24, 2003.

433. T. Kondro, A. Ishiguro, Y. Wantanabe and Y. Uchikawa. Evolutionary Construction of an immune network based behavior arbitration mechanism for autonomous mobile robots. Electrical Engineering in Japan, 123(3), pp. 1–10, 1998.

434. A. P. Kosoresow and S. A. Hofmeyr. Intrusion Detection via System Call Traces. IEEE Software, Vol. 14, No. 5, September–October 1997.

435. M. Krautmacher and W. Diger. AIS Based Robot Navigation in a Rescue Scenario. Published in the proceedings of the Third International Conference on Artificial Immune Systems (ICARIS), Catania, Italy. September 13–16, 2004.

436. K. Krishnakumar and J. Neidhoefer. Immunized Adaptive Critic for an Autonomous Aircraft Control Application. Chapter 20 in the book entitled Artificial Immune Systems and Their Applications, Publisher: Springer-Verlag, Inc., pp. 221–240, January 1999.

437. K. KrishnaKumar and J. C. Neidhoefer. Immunized Neurocontrol, Expert Systems with Applications, Vol. 13, No. 3, pp. 201–214, 1997.

438. K. KrishnaKumar and J. Neidhoefer. Immunized Adaptive Critics. ICNN, Houston, TX. June 1997.

439. K. KrishnaKumar, A. Satyadas and J. C. Neidhoefer. An immune system framework for integrating computational intelligence paradigms. In Computational Intelligence, A Dynamic Perspective, IEEE Press, 1995.

440. K. KrishnaKumar. Immunized Neurocontrol: Concepts and Initial Results, International workshop on combinations of genetic algorithms and neural networks (COGANN), IEEE Press, pp. 146–168, 1992.

441. Krohling, Zhou & Tyrrell. Evolving FPGA-based robot controllers using an evolutionary algorithm. Published in the proceedings of the 1st International Conference on Artificial Immune Systems (ICARIS), University of Kent at Canterbury, U.K., September 9–11, 2002.

442. N. Kubota, K. Shimojima and T. Fukuda. The Role of Virus Infection in Virus-evolutionary Genetic Algorithm. Published in the proceedings of IEEE International Conference on Evolutionary Computation, Nago IEEE. Japan, pp. 182–187, 1996.

L

443. Milad Lagevardi, Joseph Lewis. Artificial Immune System for Discovering Heuristics in Othello. Published in the proceedings of Genetic and Evolutionary Computation Conference GECCO 2006.

444. P. K. Lala, B. K. Kumar. Human Immune System inspired Architecture for Self-Healing Digital Systems. International Symposium on Quality Electronic Design. San Jose, California. March 18–21, 2002.

445. A. B. Lambert, R. L. King, S. H. Russ and D. S. Reese. Adaptive Analysis for the Design of Hardware Agents Using the Artificial Immune System Model for Resource Management of Heterogeneous Systems, Miss. State Technical Report No. MSSU–COE–ERC–98–10, August, 1998.

446. Gary Lamont, Mark Esslinger, Robert Ewing and Hoda Abdel-Aty-Zohdy. Evolutionary Computation in Bioinformatics and Computational Biology: An Artificial Immune System Strategy for Robust Chemical Spectra Classification via Distributed Heterogeneous Sensors. Published in the proceedings of the Congress on Evolutionary Computation (CEC). Portland, Oregon USA, June 20–23, 2004.

447. G. B. Lamont, R. E. Marmelstein and D. A. Van Veldhuizen. A distributed Architecture for a self-Adaptive Computer Virus Immune System. A chapter in the book "New Ideas in Optimization" pp. 167–183. McGraw-Hill, 1999.

448. Larisa B. Goncharova, Jacques Y. Martin-Vide C. Tarakanov A. O., Jonathan I. Timmis. Biomolecular Immune-Computer: Theoretical Basis and Experimental Simulator. Published in the proceedings of ICARIS, 4th International Conference on Artificial Immune Systems, Banff, Canada, 2005.

449. Henry Y. K. Lau, Albert Ko. Robotics, Control and Electronics: An Immuno Robotic System for Humanitarian Search and Rescue. In the proceedings of 6th international conference on Artificial Immune systems, 26th–29th August, 2007 in Santos/SP, Brazil.

450. Henry Y. K. Lau and Eugene Y. C. Wong. An AIS-Based Dynamic Routing (AISDR) Framework. Published in the proceedings of ICARIS, 4th International Conference on Artificial Immune Systems, Banff, Canada.

451. H. Y. K. Lau and V. W. K. Wong. Immunologic Responses Manipulation of AIS Agents (Conceptual Paper). Published in the proceedings of the Third International Conference on Artificial Immune Systems (ICARIS), Catania, Italy. September 13–16, 2004.

452. Henry Y. K. Lau and Vicky W. K. Wong. Immunologic Control Framework for Automated Material Handling. Published in the proceedings of Second International Conference on Artificial Immune Systems (ICARIS), September 1–3, 2003, Napier University, Edinburgh, U.K.

453. Nicholas Lay and Iain Bate. Applying Artificial Immune Systems to Real-Time Embedded Systems. In the proceedings of Congress on Evolutionary Computation (CEC) 2007, 25–28 September, Singapore.

454. Jongan Lee, Mootaek Roh, Jinseong Lee, Doheon Lee. Robotics, Control and Electronics: Clonal Selection Algorithms for 6-DOF PID Control of Autonomous Underwater Vehicles. In the proceedings of 6th international conference on Artificial Immune systems, 26th–29th August, 2007 in Santos/SP, Brazil.

455. Doheon Lee, Jungja Kim, Mina Jeong, Yonggwan Won, Seon Hee Park and Kwang-Hyung Lee. Immune-Based Framework for Exploratory Bio-Information Retrieval from the Semantic Web. Published in the proceedings of Second International Conference on Artificial Immune Systems (ICARIS), September 1–3, 2003, Napier University, Edinburgh, U.K.

456. D. Lee and H. Jun and K. Sim. Artificial Immune System for realization of co-operative strategies and group behavior in collective autonomous mobile robots. Published in the proceedings of Fourth International Symposium on Artificial Life and Robotics, Pages 232–235, AAAI. 1999.

457. D. Lee and K. Sim. Artificial Immune Network based cooperative control in collective autonomous mobile robots. Published in the proceedings of IEEE International Workshop on robot and Human Communication. Sendai, Japan, IEEE. pp. 58–63. 1997.

458. W. Lee and S. J. Stolfo. Learning Patterns from Unix Process Execution Traces for Intrusion Detection. Published in the proceedings of the AAAI workshop on AI methods in Fraud and risk management, 1997.

459. W. Lee Dong. Information-Theoretic Measures for Anomaly Detection. Citeseer. nj.nec.com/408421.html

460. Martin Lehmann and Werner Dilger. Controlling the Heating System of an Intelligent Home with an Artificial Immune System. Published in the proceedings of the 5th International Conference on Artificial Immune Systems (ICARIS), Portugal, 4–6 September, 2006.

461. W. Lei, P. Jin and J. Li-cheng. The Immune Algorithm. Acta Electronica Sinica. 28(7): pp. 74–78, 2000.

462. Wang Lei and Beat Hirsbrunner. An Evolutionary Algorithm with Population Immunity and Its Application on Autonomous Robot Control. A conceptual paper published at the 2003 IEEE Congress on Evolutionary Computation, Canberra, Australia. December 8–12, 2003.

463. Wang Lei, B. Hirsbrunner. Immune mechanism based computer security design. Published in the proceedings of International Conference on Machine Learning and Cybernetics, Vol. 4, pp. 1887–1893. November 4–5, 2002.

464. Kevin Leung, France Cheong and Christopher Cheong. Consumer Credit Scoring Using an Artificial Immune System Algorithm. In the proceedings of Congress on Evolutionary Computation (CEC) 2007, 25–28 September, Singapore.

465. Li Maojun, Tang Zhong. An artificial immune algorithm based on bidding of power market. Published in the proceedings of Power System Technology International Conference (PowerCon 2002), Vol. 4, pp. 2405–2408. October 13–17, 2002.

466. Y. Li and Jiao L. Quantum-Inspired Immune Clonal Algorithm. Published in the proceedings of ICARIS, 4th International Conference on Artificial Immune Systems, Banff, Canada, 2005.

467. Liu Shulin, Zhang Jiazhong, Shi Wengang, Huang Wenhu. Negative-selection algorithm based approach for fault diagnosis of rotary machinery. Published in the proceedings of American Control Conference, Vol. 5, pp. 3955–3960. 8–10 May 8–10, 2002.

468. G. Luh and W. Liu. Reactive immune network based mobile robot navigation. Published in the proceedings of the Third International Conference on Artificial Immune Systems (ICARIS), Catania, Italy. September 13–16, 2004.

469. G. Luh, C. Wu and W. Cheng. Artificial Immune Regulation (AIR) for Model-Based Fault Diagnosis. Published in the proceedings of the Third International Conference on Artificial Immune Systems (ICARIS), Catania, Italy. September 13–16, 2004.

470. C. Lundegaard, M. Nielsen, K. Lamberth, P. Worming, C. Sylvester-Hvid, S. Buus, S. Brunak and O. Lund. Published in the proceedings of the Third International Conference on Artificial Immune Systems (ICARIS), Catania, Italy. September 13–16, 2004.

471. Wenjian Luo, Xin Wang, Xufa Wang. Applications and Negative Selection: A Novel Fast Negative Selection Algorithm Enhanced by State Graphs. In the proceedings of 6th international conference on Artificial Immune systems, 26th–29th August, 2007 in Santos/SP, Brazil.

M

472. Zhongli Ma and Hongda Liu. Pipeline Defect Detection and Sizing Based on MFL Data Using Immune RBF Neural Networks. In the proceedings of Congress on Evolutionary Computation (CEC) 2007, 25–28 September, Singapore.

473. P. Marrack and J. W. Kappler. How the immune system recognizes the body. Scientific American, 269(3): pp. 81–89, September 1993.

474. Marwah & Boggess. Artificial Immune Systems for Classification: Some Issues.1st International Conference on Artificial Immune Systems (ICARIS), University of Kent at Canterbury, U.K., September 9–11, 2002.

475. R. E. Marmelstein, D. A. Van Veldhuizen, P. K. Harmer and G. B. Laymont. A white paper on modeling and analysis of computer immune systems using evolutionary algorithms. TR 1. Air Force Institute of Technology. WPAFB. OH. December 1999.

476. R. E. Marmelstein, D. A. Van Veldhvizen and G. B. Lamont. A Distributed Architecture for an Adaptive Computer Virus Immune. In the IEEE International Conference on Systems, Man, and Cybernetics, San Diego, October 1998.

477. K. Mathias and J. Byassee. Agent Support of Genetic Search in an Immunological Model of Sparse Distributed Memory (AAAA). Published in the proceedings of the International Conference Genetic and Evolutionary Computation (GECCO), New York, July 9–13, 2002.

478. Peter May, Jon Timmis, Keith Mander. General Applications: Immune and Evolutionary Approaches to Software Mutation Testing. In the proceedings of 6th international conference on Artificial Immune systems, 26th–29th August, 2007 in Santos/SP, Brazil.

479. Peter May, Keith Mander and Jon Timmis. Software Vaccination: An Artificial Immune System Approach to Mutation Testing. Published in the proceedings of Second International Conference on Artificial Immune Systems (ICARIS), September 1–3, 2003, Napier University, Edinburgh, U.K.

480. Nauman Mazhar, Muddassar Farooq. General Applications: BeeAIS: Artificial Immune System Security for Nature Inspired, MANET Routing Protocol, Bee-AdHoc**. In the proceedings of 6th international conference on Artificial Immune systems, 26th–29th August, 2007 in Santos/SP, Brazil.

481. D. McCoy and V. Devarajan. Artificial Immune Systems for Aerial Image Segmentation. Published in the proceedings of the IEEE International Conference on Systems, Man, and Cybernetics, Orlando, Florida, October 13, 1997.

482. Chris McEwan, Emma Hart, Ben Paechter. Modeling: Revisiting the Central and Peripheral Immune System. In the proceedings of 6th international conference on Artificial Immune systems, 26th–29th August, 2007 in Santos/SP, Brazil.

483. M. Meier-Schellersheim. Understanding information processing in the Immune System; Computer Modeling and Simulations. Acoustics, Speech, and Signal Processing. Published in the proceedings of IEEE International Conference (ICASSP). Vol. 4, pp. 4036–4039, May 13–17, 2002.

484. M. Mendao, Jon Timmis, Paul Andrews and Matthew Davies. The immune system in pieces: Computational lessons from degeneracy in the immune system. To appear in the proceedings of the First IEEE Symposium on Foundations of Computational Intelligence (FOCI) 1–5 April 2007, Honolulu, Hawaii, USA.

485. Ke Meng, Rui Xia, Ting Ji and Feng Qian. Electricity Reference Price Forecasting with Fuzzy C-means and Immune Algorithm. In the proceedings of Congress on Evolutionary Computation (CEC) 2007, 25–28 September, Singapore.

486. Filippo Menolascina, Roberto Teixeira Alves, Stefania Tommasi, Patrizia Chiarappa, Myriam Delgado, Giuseppe Mastronardi, Angelo Paradiso, Alex Freitas, Vitoantonio Bevilacqua. Induction of fuzzy rules with artificial immune systems in acgh based er status breast cancer characterization. Published in the Proceedings of the 9th annual conference on Genetic and evolutionary computation (GECCO) 2007, pp. 431–431, London, England.

487. H. Meshref, H. VanLandingham. Immune network simulation of reactive control of a robot arm manipulator. Published in the proceedings of the 2001 IEEE Mountain Workshop on Soft Computing in Industrial Applications, SMCia/01, pp. 81–85. June 25–27, 2001.

488. S. R. Michaud, G. Lemont, J. B. Zydallis, P. K. Harmer and R. Pachter. Protein Structure Prediction and Immunological EA Computation. Published in the proceedings of Genetic and Evolutionary Computation Conference (GECCO) 2001.

489. R. Michelan and F. J. Von Zuben. Decentralized Control System for Autonomous Navigation based on an Evolved Artificial Immune Network. Published in the proceedings of the special sessions on artificial immune systems in Congress on Evolutionary Computation, IEEE World Congress on Computational Intelligence, Honolulu, Hawaii, May 2002.

490. N. Mitsumoto, T. Fukuda, F. Arai and Ishihara. Control of distributed autonomous robotic system based on the biologically inspired immunological architecture. Published in the proceedings of IEEE International Conference on Robotics and Automation. Albuquerque, USA, IEEE. pp. 3551–3556, 1997.

491. N. Mitsumoto, T. Fukuda and T. Idogaki. Self-Organizing multiple robotic systems. Published in the proceedings of IEEE International Conference on Robotics and Automation. Minneapolis, USA, IEEE, pp. 1614–1619. 1996.

492. N. Mitsumoto et al. Micro Autonomous Robotic System and Biologically Inspired Immune Swarm Strategy as a Multi Agent Robotic System. Published in the proceedings of the Int. Conf. On Robotics and Automation, pp. 2187–2192, 1995.

493. P. H. Mohr, N. Ryan and J. Timmis. Exploiting Immunological Properties for Ubiquitous Computing Systems (Conceptual Paper). Published in the proceedings of the Third International Conference on Artificial Immune Systems (ICARIS), Catania, Italy. September 13–16, 2004.

494. R. R. Mohler, C. Bruni, and A. Candolfi. A System Approach to Immunology. Published in the proceedings of the IEEE, 68(8): pp. 964–990, 1980.

495. D. Morawietz, D. Chowdhury, S. Vollmar and D. Stauffer. Simulation of the kinetics of the Widom model of microemulsion. Physica A (Elsevier), Vol.187, 126, 1992.

496. K. Mori, K. Abe, M. Tsukiyama and T. Fukuda. Artificial Immune System based on Petri Nets and its Application to Production Management Systems. Published in the proceedings of Genetic and Evolutionary Computation Conference (GECCO), Las Vegas, Nevada, USA, July 8, 2000.

497. K. Mori, M. Tsukiyama and T. Fukuda. Adaptive Scheduling System Inspired by Immune System. In IEEE Int. Conf. on Systems, Man, and Cybernetics, San Diego, 1998.

498. K. Mori, M. Tsukiyama and T. Pukuda. Application of an Immune Algorithm to Multi-Optimization Problems. The 7V-ans. of the Institute of Electrical Engineers of Japan, Vol. 117-C, No.5, pp. 593–598 (in Japanese), 1997.

499. K. Mori, M. Tsukiyama and T. Fukuda. Artificial Immunity Based Management System for a Semiconductor Production Line. In 1997 IEEE Int. Conf. on Systems, Man, and Cybernetics, Vol. 1, pp. 852–856, 1997.

500. K. Mori, M. Tsukiyama and T. Fukuda. Multi-Optimization by Immune Algorithm with Diversity and Learning. 2nd Int. Conf. on Multi-Agent Systems, Workshop Notes on Immunity-Based Systems, pp. 118–123, 1996.

501. K. Mori, M. Tsukiyama and T. Fukuda. Immune Algorithm and its Application to Factory Load Dispatching Planning. 1994 JAPAN-U.S.A. Symposium on Flexible Automation, pp. 1343–1346, 1994.

502. K. Mori, M. Tsukiyarna and T. Fukuda. Load Dispatching Planning by Immune Algorithm with Diversity and Learning. 7th Int. Conf. on Systems Research, Informatics and Cybernetics, Vol. 11, pp. 136–141, 1994.

503. K. Mori, M. Tsukiyama and T. Fukuda. Immune Algorithm with Searching Diversity and its Application to Resource Allocation Problem. The Trans. of the Institute of Electrical Engineers of Japan, Vol. 113-C, No. 10, pp. 872–878 (in Japanese), 1993.

504. Morrison & Aickelin. An Artificial Immune System as a Recommender for Web Sites. In 1st International Conference on Artificial Immune Systems (ICARIS), University of Kent at Canterbury, U.K., September 9–11, 2002.

505. M. Saniee Abadeh, J. Habibi, M. Daneshi, M. Jalali and M. Khezrzadeh. Intrusion Detection Using a Hybridization of Evolutionary Fuzzy Systems and Artificial Immune Systems. In the proceedings of Congress on Evolutionary Computation (CEC) 2007, 25–28 September, Singapore.

506. M. Zubair Shafiq, Muddassar Farooq. General Applications: Defence Against 802.11 DoS Attacks Using Artificial Immune System. In the proceedings of 6th international conference on Artificial Immune systems, 26th–29th August, 2007 in Santos/SP, Brazil.

507. M. Zubair Shafiq, Mehrin Kiani, Bisma Hashmi and Muddassar Farooq. Extended thymus action for reducing false positives in ais based network intrusion detection systems. Published in the Proceedings of the 9th annual conference on Genetic and evolutionary computation (GECCO) 2007, pp. 182–182, London, England.

N

508. D. Na and Lee D. Mathematical Modeling of Immune Suppression. Published in the proceedings of ICARIS, 4th International Conference on Artificial Immune Systems, Banff, Canada, 2005.

509. D. Na, I. Park, K.H. Lee and D. Lee. Integration of Immune Models Using Petri Nets. Published in the proceedings of the Third International Conference on Artificial Immune Systems (ICARIS), Catania, Italy. September 13–16, 2004.

510. Nikolaos Nanas, Anne De Roeck. Search and Optimization: Multimodal Dynamic Optimisation: From Evolutionary Algorithms to Artificial Immune Systems. In the proceedings of 6th international conference on Artificial Immune systems, 26th–29th August, 2007 in Santos/SP, Brazil.

511. Nikolaos Nanas, Anne de Roeck and Victoria Uren. Immune-Inspired Adaptive Information Filtering. Published in the proceedings of the 5th International Conference on Artificial Immune Systems (ICARIS), Portugal, 4–6 September, 2006.

512. N. Nanas, V. Uren and A. de Roeck. Nootropia: a User Profiling Model Based on a Self-Organizing Term Network (Conceptual Paper). Published in the proceedings of the Third International Conference on Artificial Immune Systems (ICARIS), Catania, Italy. September 13–16, 2004.

513. P. Narasimhan, K. P. Kihlstrom, L. E. Moser, P. M. Mellliar-Smith. Providing Support for Survivable CORBA Applications with the Immune System. 19th IEEE International Conference on Distributed Computing Systems. Austin, Texas. May 31–June 04, 1999.

514. Nareli Cruz-Cortés, Daniel Trejo-Pérez, Carlos A. Coello Coello. Handling Constraints in Global Optimization Using an Artificial Immune System. Published in the proceedings of ICARIS, 4th International Conference on Artificial Immune Systems, Banff, Canada, 2005.

515. O. Nasraoui, F. González, C. Cardona, C. Rojas and D. Dasgupta. 'A Scalable Artificial Immune System Model for Dynamic Unsupervised Learning'. Published in the proceedings of the Genetic and Evolutionary Computation Conference (GECCO), July 12–16, 2003. LNCS 2723, p. 219 ff.

516. O. Nasraoui, D. Dasgupta and F. Gonzalez. An Novel Artificial Immune System Approach to Robust Data Mining. Published in the proceedings of the International Conference Genetic and Evolutionary Computation (GECCO), New York, July 9–13, 2002.

517. O. Nasraoui, F. Gonzalez and D. Dasgupta. The Fuzzy Artificial Immune System: Motivations, Basic Concepts, and Application to Clustering and Web Profiling. Published at IEEE International Conference on Fuzzy Systems. Published in the proceedings of the IEEE World Congress on Computational Intelligence, Hawaii, May 12–17, 2002.

518. O. Nasraoui, D. Dasgupta and F. Gonzalez. The Promise and Challenges of Artificial Immune System Based Web Usage Mining: Preliminary Results. Presented at the workshop on Web Analytics at Second SIAM International Conference on Data Mining (SDM), Arlington, VA, April 11–13, 2002.

519. Mark Neal, Jan Feyereisl, Rosario Rascuna and Xiaolei Wang. Don't Touch Me, I'm Fine: Robot Autonomy Using An Artificial Innate Immune System. Published in the proceedings of the 5th International Conference on Artificial Immune Systems (ICARIS), Portugal, 4–6 September, 2006.

520. Mark Neal. Meta-Stable Memory in an Artificial Immune Network. Published in the proceedings of Second International Conference on Artificial Immune Systems (ICARIS), September 1–3, 2003, Napier University, Edinburgh, U.K.

521. M. Neal. An Artificial Immune System for Continuous Analysis of Time-Varying Data. In 1st International Conference on Artificial Immune Systems (ICARIS), University of Kent at Canterbury, U.K., September 9–11, 2002.

522. M. Neal, J. Hunt and J. Timmis. Augmenting an artificial immune network. Published in the proceedings of Int. Conf. Systems and Man and Cybernetics, IEEE, pp. 3821–3826, San Diego, California, U.S.A., 1998.

523. J. Newborough and Stepney S. A Generic Framework for Population-Based Algorithms, Implemented on Multiple FPGAs. Published in the proceedings of ICARIS, 4th International Conference on Artificial Immune Systems, Banff, Canada, 2005.

524. Giuseppe Nicosia, Vincenzo Cutello and Mario Pavone. An Immune Algorithm with Hyper-Macromutations for the 2D Hydrophilic-Hydrophobic Model. Published in the proceedings of the Congress on Evolutionary Computation (CEC). Portland, Oregon USA, June 20–23, 2004.

525. Giuseppe Nicosia and Alexander Tarakanov. Foundations of immunocomputing for intelligent signal processing. To appear in the proceedings of the First IEEE Symposium on Foundations of Computational Intelligence (FOCI) 1–5 April 2007, Honolulu, Hawaii, USA.

526. N. Nikolaev, H. Iba and V. Slavov. Inductive Genetic Programming with Immune Network Dynamics. In: L. Spector, W. B. Langdon, U.-M. O'Reilly and P. J. Angeline (Eds.), Advances in Genetic Programming 3, Chapter 15, MIT Press, Cambridge, MA, pp. 355–376. (1999).

527. Nikolaos D. Atreas, Costas G. Karanikas and Alexander Tarakanov: Signal Processing by an Immune Type Tree Transform. In 2nd International Conference on Artificial Immune Systems, Edinburgh, U.K., September 1–3, 2003.

528. F. Niño, D. Gómez, and R. Vejar. A Novel Immune Anomaly Detection Technique Based on Negative Selection. Published in the proceedings of the Genetic and Evolutionary Computation Conference (GECCO) [Poster], Chicago, IL, USA, July 12–16, 2003. LNCS 2723, p. 243 ff.

529. F. Nino and O. Beltran. A change detection software agent based on immune mixed selection. Published in the proceedings of the special sessions on artificial immune systems in Congress on Evolutionary Computation, IEEE World Congress on Computational Intelligence, Honolulu, Hawaii, May 2002.

530. H. Nishiyama, F. Mizoguchi. Design of Security System Based on Immune System. Tenth IEEE International Workshops on Enabling Technologies: Infrastructure for Collaborative Enterprises. Massachusetts. June 20–22, 2001.

531. A. J. Noest, K. Takumi and R. de Boer, Pattern formation in B-cell immune network: Domains and dots in shape-space. Physica D 105:285–306, 1997.

532. Ian Nunn and Tony White. The Application of Antigenic Search Techniques to Time Series Forecasting. Published in the proceedings of the Genetic and Evolutionary Computation Conference (GECCO), Washington, D.C., June 25–29 2005.

O

533. Robert Oates, Julie Greensmith, Uwe Aickelin, Jonathan Garibaldi, Graham Kendall. Robotics, Control and Electronics: The Application of a Dendritic Cell Algorithm to a Robotic Classifier. In the proceedings of 6th international conference on Artificial Immune systems, 26th–29th August, 2007 in Santos/SP, Brazil.

534. Terri Oda and White T. Immunity from Spam: An Analysis of an Artificial Immune System for Junk Email Detection. Published in the proceedings of ICARIS, 4th International Conference on Artificial Immune Systems, Banff, Canada, 2005.

535. Terri Oda and Tony White, "Spam Detection using an Artificial Immune System." ACM Crossroads Magazine, Winter 2004. Accepted August 2004 (not published).

536. Terri Oda and Tony White, "Increasing the Accuracy of a Spam-Detecting Artificial Immune System." In *Proceedings of the Congress on Evolutionary Computation (CEC 2003)*, Canberra, Australia, December 2003. Proceedings volume 1:390–396.

537. Terri Oda and Tony White. Developing an Immunity to Spam. Published in the proceedings of the Genetic and Evolutionary Computation Conference (GECCO), Chicago, IL, USA, July 12–16, 2003. LNCS 2723, p. 231 ff.

538. Z. X. Ong, J. C. Tay, C. K. Kwoh. Applying the Clonal Selection Principle to Find Flexible Job-Shop Schedules. Published in the proceedings of ICARIS, 4th International Conference on Artificial Immune Systems, Banff, Canada, 2005.

539. T. Okamoto and Y. Ishida. A distributed approach against computer virus inspired by the immune system. IEICE Transactions on Communications, Tokyo. E83-B (5): 908–915. 2000.

540. Fabricio Olivetti de Franca, Fernando J. Von Zuben and Leandro Nunes de Castro. An Artificial Immune Network for Multimodal Function Optimization on Dynamic Environments. Published in the proceedings of the Genetic and Evolutionary Computation Conference (GECCO), Washington, D.C., June 25–29 2005.

541. M. Oprea and S. Forrest. How the immune system generates diversity: Pathogen space coverage with random and evolved antibody libraries. 1999 Genetic and Evolutionary Computation Conference (GECCO), July 1999.

542. M. Oprea and S. Forrest. Simulated evolution of antibody gene libraries under pathogen selection. In IEEE Int. Conf. on Systems, Man, and Cybernetics, San Diego, 1998.

543. M. Ostaszewski, F. Seredynski, P. Bouvry. Immune Anomaly Detection Enhanced with Evolutionary Paradigms. Published in the proceedings of the conference on Genetic and evolutionary computation (GECCO), July 8–12, 2006.

544. Nick D. Owens, Jon Timmis, Andrew J. Greensted, Andy M. Tyrell. Robotics, Control and Electronics: On Immune Inspired Homeostasis for Electronic Systems. In the proceedings of 6th international conference on Artificial Immune systems, 26th–29th August, 2007 in Santos/SP, Brazil.

P

545. Rodrigo Pasti, Leandro Nunes de Castro. Classification and Clustering: The Influence of Diversity in an Immune-Based Algorithm to Train MLP Networks. In the proceedings of 6th international conference on Artificial Immune systems, 26th–29th August, 2007 in Santos/SP, Brazil.

546. Rodrigo Pasti and Leandro de Castro. An Immune and a Gradient-based Method to Train Multi-layer Perceptron Neural Networks. Published in the proceedings of IEEE World Congress on Computational Intelligence/Congress on Evolutionary Computation, Vancouver, Canada, July 16–21, 2006.

547. Rodrigo Pasti and Leandro de Castro. A Neuro-Immune Network for Solving the Traveling Salesman Problem. Published in the proceedings of IEEE World Congress on Computational Intelligence/Congress on Evolutionary Computation, Vancouver, Canada, July 16–21, 2006.

548. Rafal Pasek. Theoretical basis of Novelty Detection in Time Series using Negative Selection Algorithms. Published in the proceedings of the 5th International Conference on Artificial Immune Systems (ICARIS), Portugal, 4–6 September, 2006.

549. Andreas Pietzowski, Benjamin Satzger, Wolfgang Trumler, Theo Ungerer. An Artificial Immune System and its Integration into an Organic Middleware for Self-Protection. Published in the proceedings (as Poster) of the Genetic and Evolutionary Computation Conference (GECCO), Seattle, Washington, July 8–12, 2006.

550. W. E. Paul. (Ed.) Immunology: Recognition and Response. Readings from Scientific American. New York: W. H. Freeman and Company, 1991.

551. Fabricio de Paula, Leandro de Castro and Paulo de Geus. An Intrusion Detection System Using Ideas from the Immune System. Published in the proceedings of the Congress on Evolutionary Computation (CEC). Portland, Oregon USA, June 20–23, 2004.

552. F. S. Paula, M. A. Reis, Fernandes, D. A. M. Geus, P. L. Adenoids: A hibrid IDS based on the immune system. Published in the proceedings of the 9th International Conference on Neural Information Processing, ICONIP, Vol. 3, pp. 479–1484. November 18–22, 2002.

553. J. K. Percus, O. Percus and A. S. Person. Predicting the size of the antibody-combining region from consideration of efficient self/non-self discrimination. Published in the proceedings of the National Academy of Science, 60: pp. 1691–1695, 1993.

554. J. K. Percus, O. Percus, and A. S. Person. Probability of Self Non-self discrimination. In A. S. Perelson and G. Weisbuch, (eds.), Theoretical and Experimental Insights into Immunology, pp. 63–70. Springer-Verlag, 1992.

555. A. S. Perelson and F. W. Weigel. Some Design Principles for Immune System Recognition. In the Journal Complexity, John Wiley & Sons, Inc, Vol. 4, No. 5, 1999.

556. A. S. Perelson and G. Weisbuch. Immunology for physicists. Review of Modern Physics, 69:1219–1265. June 1997.

557. A. S. Perelson, R. Hightower and S. Forrest. Evolution (and learning) of V-region genes. Research in Immunology Vol. 147, pp. 202–208, 1996.

558. A. S. Perelson and G. Weisbuch. Eds. Theoretical and Experimental insights into immunology, chapter Probability of Self-Nonself discrimination, pp. 63–70. Springer-Verlag. NY, 1992.

559. A. S. Perelson. Immune network theory. Immunological Reviews, (10): 5–36, 1989.

560. A. S. Perelson and G. F. Oster. Theoretical studies of clonal selection: Minimal antibody repertoire size and reliability of self- non-self discrimination. J. Theoret. Biol., 81:645–670, 1979.

561. Kemal Polat, Sadik Kara, Fatma Latifoglu and Salih Gunes. Approach to Resource Allocation Mechanism in Artificial Immune Recognition System: Fuzzy Resource Allocation Mechanism and Application to Diagnosis of Atherosclerosis Disease. Published in the proceedings of the 5th International Conference on Artificial Immune Systems (ICARIS), Portugal, 4–6 September, 2006.

562. M. A. Potter and K. A. De Jong. The Coevolution of Antibodies for Concept Learning. Published in the proceedings of the Parallel Problem Solving from Nature (PPSN), Amsterdam, 1998.

563. S. Pramanik, R. Kozma and D. Dasgupta. Simulation of Germinal Center Dynamics using Cascaded Hopfield Neural Networks. Technical Report CS-02-002, May 2002.

564. S. Pramanik, R. Kozma and D. Dasgupta. Dynamical Neuro-Representation of an Immune Model and its Application for Data Classification. Published in the proceedings of IJCNN, WCCI, May 2002.

565. C. Pu, A. Black, C. Cowan and J. Walpole. A Specialization Toolkit to Increase the Diversity in Operating Systems. Presented at ICMAS Workshop on Immunity-Based Systems, December 10, 1996.

R

566. T. K. Rahaman. Artificial Immune-Based Optimization Technique For Solving Economic Dispatch In Power System. Published in the proceedings of International Workshop on Natural and Artificial Immune Systems (NAIS) Vietri sul Mare, Salerno, Italy, June 9–10, 2005.

567. Pedro A. Reche and Ellis L. Reinherz. Definition of MHC peptide binding repertoires. Published in the proceedings of the Third International Conference on Artificial Immune Systems (ICARIS), Catania, Italy. September 13–16, 2004.

568. Maria-Cristina Riff and Marcos Zuniga. Towards an Immune System that Solves CSP. In the proceedings of Congress on Evolutionary Computation (CEC) 2007, 25–28 September, Singapore.

569. M. J. Robbins and S. M. Garrett. Evaluating Theories of Immunological Memory Using Large-Scale Simulations. Published in the proceedings of ICARIS, 4th International Conference on Artificial Immune Systems, Banff, Canada, 2005.

570. I. Roitt. Essential Immunology. Ninth Edition. Pub. Blackwell Science. Specific Acquired Immunity. pp. 22–39, 1997.

571. I. Roitt. Essential Immunology. Ninth Edition. Pub. Blackwell Science. Ontogeny and Phylogeny. pp. 223–250, 1997.

572. Romero, Fernando Nino. Keyword extraction using an artificial immune system. Published in the Proceedings of the 9th annual conference on Genetic and evolutionary computation (GECCO) 2007, pp. 181–181, London, England.

573. Diego Romero and Fernando Nino. An Immune-based Multilayered Cognitive Model for Autonomous Navigation. Published in the proceedings of IEEE World Congress on Computational Intelligence (special session on Evolutionary Intelligent Agents) in Congress on Evolutionary Computation, Vancouver, Canada, July 16–21, 2006.

574. G. W. Rowe. The Theoretical Models in Biology. Oxford University Press, first edition, 1994.

575. P. K. Roy, R. Kozma and D. Dutta Majumder. From Neurocomputation to Immunocomputation: A model and algorithm for fluctuation induced stability and phase transitions in biological systems. In the Special Issue on Artificial Immune Systems of the journal IEEE Transactions on Evolutionary Computation, Vol. 6, No. 3, June 2002.

S

576. S. Şahan, Polat K, Kodaz H, Güneş S. The Medical Applications of Attribute Weighted Artificial Immune System (AWAIS): Diagnosis of Heart and Diabetes Diseases. Published in the proceedings of ICARIS, 4th International Conference on Artificial Immune Systems, Banff, Canada, 2005.

577. S. Sahan and G.P.S. Raghava. BcePred: Prediction of Continuous B-Cell epitopes in antigenic sequences using physico-chemical properties. Published in the proceedings of the Third International Conference on Artificial Immune Systems (ICARIS), Catania, Italy. September 13–16, 2004.

578. R. M. Z. Santos and A. T. Bernardes. The stable-chaotic transition on cellular automata used to model the immune repertoire. Physica *A*, 219: pp. 1–12, 1995.

579. I. Safro and L. A. Segel, Collective versions of playable games as metaphors for complex biosystems: team collect four. Complexity, 2003.

580. S. Sarafijanovic and J. Le Boudec. An Artificial Immune System for Misbehavior Detection in Mobile Ad-Hoc Networks with Virtual Thymus, Clustering, Danger Signal and Memory Detectors. Published in the proceedings of the Third International Conference on Artificial Immune Systems (ICARIS), Catania, Italy. September 13–16, 2004.

581. S. Sathyanath, F. Sahin. Application of artificial immune system based intelligent multi agent model to a mine detection problem. Published in the proceedings of Systems, Man and Cybernetics, IEEE International Conference, pp. 6, Vol. 3. October 6–9, 2002.

582. Sathyanath & Sahin. AISIMAM—An Artificial Immune System Based Intelligent Multi-Agent Model and its Application to a Mine Detection Problem. In 1st International Conference on Artificial Immune Systems (ICARIS), University of Kent at Canterbury, U.K., September 9th–11th, 2002.

583. S. Sathyanath and F. Sahin. An AIS Approach to a color image classification problem in a real time industrial application. Published in the proceedings of the IEEE systems, man and cybernetics conference. 2001.

584. Andrew Secker, Alex A. Freitas and Jon Timmis. A Danger Theory Inspired Approach to Web Mining. Published in the proceedings of Second International Conference on Artificial Immune Systems (ICARIS), September 1–3, 2003, Napier University, Edinburgh, U.K.

585. Stefan Schadwinkel, Werner Dilger. A Dynamic Approach to Artificial Immune Systems utilizing Neural Networks. Published in the proceedings (as Poster) of the Genetic and Evolutionary Computation Conference (GECCO), Seattle, Washington, July 8–12, 2006.

586. Holger Schmidtchen and Ulrich Behn - Randomly Evolving Idiotypic Networks: Analysis of Building Principles. Published in the proceedings of the 5th International Conference on Artificial Immune Systems (ICARIS), Portugal, 4–6 September, 2006.

587. L. A. Segel. Some Spatio-Temporal Models in Immunology in honor of Manuel Velarde. Bifurcation and Chaos, 2003.

588. L. A. Segel. How does the immune system see to it that it is doing a good job? Graft 4 (6): 15–18, 2001.

589. L. A. Segel and R. L. Bar-Or. On the role of feedback on promoting conflicting goals of the Adaptive Immune system. J. Immunol. 163, pp. 1342–1349, 1999.

590. L. A. Segel and R. L. Bar-Or. Immunology viewed as the Study of an Autonomous Decentralized system. Chapter 4 in the book entitled Artificial Immune Systems and Their Applications, Publisher: Springer-Verlag, Inc., pp. 65–86, January 1999.

591. Adriane B. S. Serapião, José Ricardo P. Mendes, Kazuo Miura. Classification and Clustering: Artificial Immune Systems for Classification of Petroleum Well Drilling Operations. In the proceedings of 6th international conference on Artificial Immune systems, 26th–29th August, 2007 in Santos/SP, Brazil.

592. Hamed Shah-Hosseini. The Time Adaptive Self-Organizing Map Is a Neural Network Based on Artificial Immune System. Poster presentation at Congress on Evolutionary Computation (CEC at WCCI), Vancouver (Canada), July 16–21, 2006.

593. Joseph M Shapiro, Gary B Lamont and Gilbert L Peterson. An Evolutionary Algorithm to Generate Hyper-Ellipsoid Detectors for Negative Selection. Published in the proceedings of the Genetic and Evolutionary Computation Conference (GECCO), Washington, D.C., June 25–29 2005.

594. S. P. N. Singh. Anomaly detection using negative selection based on the r-contiguous matching rule. In 1st International Conference on Artificial Immune Systems (ICARIS), University of Kent at Canterbury, U.K., September 9–11, 2002.

595. S. P. N. Singh and S. M. Thayer. A Foundation for Kilorobotic Exploration. Published in the proceedings of the special sessions on artificial immune systems in Congress on Evolutionary Computation, IEEE World Congress on Computational Intelligence, Honolulu, Hawaii, May 2002.

596. S. P. N. Singh and Scott M. Thayer. Immunology Directed Methods for Distributed Robotics: A novel, Immunity–Based Architecture for Robust Control & Coordination. Published in the proceedings of SPIE: Mobile Robots XVI, vol. 4573, November 2001.

597. Thomas Stibor. Applications and Negative Selection: Phase Transition and the Computational Complexity of Generating r-contiguous Detectors. In the proceedings of 6th international conference on Artificial Immune systems, 26th–29th August, 2007 in Santos/SP, Brazil.

598. Thomas Stibor and Rao Vemuri. An investigation on the compression quality of aiNet. To appear in the proceedings of the First IEEE Symposium on Foundations of Computational Intelligence (FOCI') 1–5 April 2007, Honolulu, Hawaii, USA.

599. Thomas Stibor, Jon Timmis and Claudia Eckert - On Permutation Masks in Hamming Negative Selection. Published in the proceedings of the 5th International Conference on Artificial Immune Systems (ICARIS), Portugal, 4–6 September, 2006.

600. T Stibor, Timmis J, Claudia Eckert. A Comparative Study of Real-Valued Negative Selection to Statistical Anomaly Detection Techniques. Published in the proceedings of ICARIS, 4th International Conference on Artificial Immune Systems, Banff, Canada, 2005.

601. T Stibor, Philipp Mohr, Jonathan Timmis and Claudia Eckert. Is Negative Selection Appropriate for Anomaly Detection? Published in the proceedings of the Genetic and Evolutionary Computation Conference (GECCO), Washington, D.C., June 25–29 2005.

602. V. A. Skormin, J. G. Delgado-Frias, Dennis L. McGee, J. V. Giordano, L. J. Popyack, V. I. Gorodetski and A. O. Tarakanov. BASIS: A Biological Approach to System Information Security. Presented at the International Workshop MMM-ACNS. St. Petersburg, Russia, May 21–23, 2001.

603. V. Slavov and N. Nikolaev. Immune Network Dynamics for Inductive Problem Solving, In: A.E. Eiben, T.Back, M.Schoenauer, and H.P. Schwefel (Eds.) Parallel Problem Solving from Nature, PPSN V, LNCS-1498, Springer, Berlin, pp. 712–721, 1998.

604. D. J. Smith. Applications of bioinformatics and computational biology to influenza surveillance and vaccine strain selection *Vaccine*, Vol 21, 1758–1761, 2003.

605. D. J. Smith, A. S. Lapedes, S. Forrest, J. C. deJong, A. D. M. E. Osterhaus, R. A. M. Fouchier, N. J. Cox, and A. S. Perelson, Modeling the effects of updating the influenza vaccine on the efficacy of repeated vaccination. In: Options for the control of influenza virus IV, eds. A.D.M.E. Osterhaus, N. Cox, and A. Hampson, Excerpta Medica, International Congress Series 1219, Amsterdam, 655–660, 2001.

606. D. J. Smith, S. Forrest, D. H. Ackley and A. S. Perelson. Variable efficacy of repeated annual influenza vaccination. Published in the proceedings of the National Academy of Sciences 96:14001–14006, 1999.

607. D. J. Smith, S. Forrest, D. H. Ackley and A. S. Perelson. Using lazy evaluation to simulate realistic-size repertoires in models of the immune system. Bulletin of Mathematical Biology Vol. 60, pp. 647–658, 1998.

608. D. J. Smith, S. Forrest, D. H. Ackley and A. S. Perelson. Modeling the effect of prior infection on vaccine efficacy. Chapter 8 in the book entitled Artificial Immune Systems and Their Applications, Publisher: Springer-Verlag, Inc., pp. 144–152, January 1999. Also presented at the 1997 IEEE International conference On Systems, man, and cybernetics. October 1997.

609. D. J. Smith, S. Forrest, R. R. Hightower and A. S. Perelson. Deriving shape-space parameters from immunological data for a model of cross-reactive memory. Journal of Theoretical Biology Vol. 189, pp. 141–150, 1997.

610. D. J. Smith, S. Forrest, and A. S. Perelson, Immunological memory is associative, Chapter 6 in the book entitled Artificial Immune Systems and Their Applications, Publisher: Springer-Verlag, Inc., pp. 105–112, January 1999. (Also presented at the Intnl. Conf. on Multiagent Systems, Workshop notes, 62–70, Kyoto, Japan, 1996).

611. D. J. Smith, Towards a Model of Associative Recall in Immunological Memory, Masters project, Computer Science Department, University of New Mexico, Technical Report, 94-9, Albuquerque NM, May 1994.

612. D. J. Smith. A Literature Review of Original Antigenic Sin, Computer Science Department, University of New Mexico, Technical Report, 94–10, Albuquerque NM, May 1994.

613. R. E. Smith, S. Forrest, and A. S. Perelson. Searching for Diverse, Cooperative Populations with Genetic Algorithms. Evolutionary Computation, Vol. 1, No. 2, pp. 127–149, 1993.

614. R. E. Smith, S. Forrest and A. S. Perelson. Population Diversity in an Immune System Model: Implication for Genetic Search. Foundation of Genetic Algorithms 2, L. D. Whitley (Ed.), Morgan Kaufmann, San Francisco, CA, pp. 153–165, 1993.

615. Ludmilla Sokolova. Index Design by Immunocomputing. Published in the proceedings of Second International Conference on Artificial Immune Systems (ICARIS), September 1–3, 2003, Napier University, Edinburgh, U.K.

616. L. Sokolova & Sokolova. Immuncomputing for Complex Interval Objects. In 1st International Conference on Artificial Immune Systems (ICARIS), University of Kent at Canterbury, U.K., September 9th–11th, 2002.

617. A. Somayaji, "Immunology, Diversity, and Homeostasis: The Past and Future of Biologically-Inspired Computer Defenses." *Information Security Technical Report (ISTR)*, August 15, 2007.

618. A. Somayaji and S. Forrest. Automated Response Using System-Call Delays. Usenix 2000.

619. A. Somayaji, S. Hofmeyr, and S. Forrest. Principles of a Computer Immune System. 1997 New Security Paradigms Workshop pp. 75–82, 1998.

620. E. H. Spafford. Computer Viruses as Artificial Life. Journal Of Artificial Life, Vol. 1, No. 3, pp. 249–pp. 265, 1994.

621. Peter Spellward and Tim Kovacs. On the Contribution of Gene Libraries to Artificial Immune Systems. Published in the proceedings of the Genetic and Evolutionary Computation Conference (GECCO), Washington, D.C., June 25–29 2005.

622. Susan Stepney, John A. Clark, Colin G. Johnson, Derek Partridge and Robert E. Smith. Artificial Immune Systems and the Grand Challenge for Non-Classical Computation. Published in the proceedings of Second International Conference on Artificial Immune Systems (ICARIS), September 1–3, 2003, Napier University, Edinburgh, U.K.

623. S. Stepney, R. E. Smith, J. Timmis and A. M. Tyrrell. Towards a Conceptual Framework for Artificial Immune Systems (Conceptual Paper). Published in the proceedings of the Third International Conference on Artificial Immune Systems (ICARIS), Catania, Italy. September 13–16, 2004.

624. Daniel Stevens, Sanjoy Das, Bala Natarajan. A multi-objective algorithm for DS-CDMA code design based on the clonal selection principle. Published in the proceedings of the conference on Genetic and evolutionary computation GECCO, June 2005 Publisher: ACM Press.

625. J. Stewart and J. Carneiro. The Central and Peripheral Immune systems: Modeling and Simulation. Chapter 3 in the book entitled Artificial Immune Systems and Their Applications, Publisher: Springer-Verlag, Inc., pp. 47–61, January 1999.

626. Thomas Stibor, Jon Timmis and Claudia Eckert - On Permutation Masks in Hamming Negative Selection. Published in the proceedings of the 5th International Conference on Artificial Immune Systems (ICARIS), Portugal, 4–6 September, 2006.

627. Thomas Stibor, Jonathan Timmis and Eckert Claudia. The Link between r-contiguous Detectors and k-CNF Satisfiability. Published in the proceedings of IEEE World Congress on Computational Intelligence (special session on recent development in artificial immune systems) in Congress on Evolutionary Computation, Vancouver, Canada, July 17–21, 2006.

628. Thomas Stibor, Kpatscha M. Bayarou. An investigation of r-chunk detector generation on higher alphabets. Published in the proceedings of International Conference on Genetic and Evolutionary Computation (GECCO). Seattle, Washington USA, June 26–30, 2004.

629. J. Suzuki and Y. Yamamoto. A Decentralized Policy Coordination Facility in OpenWebServer. Published in the proceedings of SPA 2000.

630. J. Suzuki and Y. Yamamoto. Building an Artificial Immune Network for Decentralized Policy Negotiation in a Communication Endsystem: OpenWebServer/iNexus Study. Published in the proceedings of the 4th World Multiconference on Systemics, Cybernetics and Informatics (SCI 2000).

631. J. Suzuki and Y. Yamamoto. iNet: An Extensible Framework for Simulating Immune Network. Published in the proceedings of IEEE International Conference on Systems, Man and Cybernetics (SMC), Nashville, October 8–11, 2000.

632. J. Suzuki and Y. Yamamoto. The Reflection Pattern in the Immune System.Published in the proceedings of OOPSLA '98 Workshop on Non-Software Examples of Patterns of Software Architecture.

T

633. S. A. Taheri, G. Calva. Imitating the human immune system capabilities for multi-agent federation formation. Published in the proceedings of the 2001 IEEE International Symposium on Intelligent Control, (ISIC), pp. 25–30, September 5–7, 2001.

634. Takama, Yasufumi. Visualization of Topic Distribution Based on Immune Network Model. Published in the proceedings of the Genetic and Evolutionary Computation Conference (GECCO), Chicago, IL, USA, July 2003.

635. Yasufumi Takama. Visualization of Topic Distribution Based on Immune Network Model. Published in the proceedings of the Genetic and Evolutionary Computation Conference (GECCO) [Poster], Chicago, IL, USA, July 12–16, 2003. LNCS 2723, p. 246.

636. T. Takuma, N. Saiwaki, S. Nishida, K. Shinosaki and M. Takeda. An Approach to Visualization of active position in Brain by MEG. Published in the proceedings of IEEE International Conference on Systems, Man and Cybernetics (SMC), Nashville, October 8–11, 2000.

637. K. Takahashi and Y. Yamada. Application of an immune feedback mechanism to control systems. JSME International Journal. Series C. 41(2): 184–191, 1998.

638. W. Tan and Z. Y. Stochastic. Models of Immune Response during HIV Pathogenesis Under Treatment by HAART in HIV-Infected Individuals. Published in the proceedings of Genetic and Evolutionary Computation Conference (GECCO), Las Vegas, Nevada, USA, July 8, 2000.

639. W. Tan, Z. Xiang. Estimating and predicting the number of free HIV and T Cells by Non Linear Kalman Filter. Chapter 7 in the book entitled Artificial Immune Systems and Their Applications, Publisher: Springer-Verlag, Inc., pp. 115–138, January 1999.

640. Na Tang, V. Rao Vemuri. An artificial immune system approach to document clustering Published in the proceedings of the 2005 ACM symposium on Applied computing, March 2005.

641. Z. Tang, T. Yamaguchi, K. Tashima, O. Ishizuka and K. Tanno. Multiple valued immune network model and its simulations. Published in the proceedings of the 27th International Symposium on Multiple Valued Logic, Antigonish, Canada. pp. 233–238, 1997.

642. Alexander O. Tarakanov. Spatial Formal Immune Network. Published in the proceedings of the Genetic and Evolutionary Computation Conference (GECCO) [Poster], Chicago, IL, USA, July 12–16, 2003. LNCS 2723, p. 248 f.

643. Tarakanov, Goncharova & Gupalova. Immunocomputing for Bioarrays. In 1st International Conference on Artificial Immune Systems (ICARIS), University of Kent at Canterbury, U.K., September 9th–11th, 2002.

644. A. Tarakanov and D. Dasgupta. An Immunochip Architecture and its Emulation. Published in the proceedings of the NASA/DoD Conference on Evolvable Hardware, July 15–18, 2002.

645. A. Tarakanov and V. Skormin. Pattern Recognition by Immunocomputing. Published in the proceedings of the special sessions on artificial immune systems in Congress on Evolutionary Computation, IEEE World Congress on Computational Intelligence, Honolulu, Hawaii, May 2002.

646. A. Tarakanov. Information Security with formal Immune Networks. Presented at the International Workshop MMM-ACNS. St. Petersburg, Russia, May 21–23, 2001.

647. A. Tarakanov, S. Sokolova, A. Aikimbayev and B. Abramov. Immunocomputing of the natural plague foci. Genetic and Evolutionary Computation Conference (GECCO), Las Vegas, Nevada, USA, July 8, 2000.

648. A. Tarakanov and D. Dasgupta, A formal model of an artificial immune system. In the journal BioSystems, Vol. 55/1-3, pp. 151–158, February 2000.

649. A. Tarakanov and D. Dasgupta. A Formal Immune System, Presented at the Third International Workshop on Information Processing in Cells and Tissues (IPCAT), Indianapolis, August 23–24, 1999.

650. Dan W Taylor and David W Corree. An Investigation of the Negative Selection Algorithm for Fault Detection in Refrigeration Systems. Published in the proceedings of Second International Conference on Artificial Immune Systems (ICARIS), Napier University, Edinburgh, U.K., September 1–3, 2003.

651. I. Tazawa, S. Koakustu and H. Hirata. An evolutionary optimization based on the immune system and its application to the VLSI floor plan design problem. Trans. Of the Institute of Electrical Engineers of Japan. Part C. 117-C (7): 821–828. 1997.

652. Terri Oda and Tony White. Detecting Spam Using an Artificial Immune System. Published as a Conceptual Paper at the 2003 IEEE Congress on Evolutionary Computation, Canberra, Australia, December 8th–12th, 2003.

653. Johannes Textor, Juergen Westermann. Modeling: Modelling Migration, Compartmentalization and Exit of Naive T Cells in Lymph Nodes Without Chemotaxis. In the proceedings of 6th international conference on Artificial Immune systems, 26th–29th August, 2007 in Santos/SP, Brazil.

654. S. M. Thayer, S. P. N. Singh. Development of an immunology-based multi-robot coordination algorithm for exploration and mapping domains. Published in the proceedings of Intelligent Robots and System, International Conference (IEEE/RSJ 2002), Vol. 3, pp. 2735–2739. September 30–October 5, 2002.

655. P. Tieri, S. Valensin, C. Franceschi, C. Morandi and G. C. Castellani. Memory and selectivity in evolving scale-free immune networks. Published in the proceedings of Second International Conference on Artificial Immune Systems (ICARIS), September 1–3, 2003, Napier University, Edinburgh, U.K.

656. J. Timmis. Challenges for Artificial Immune Systems. Published in the proceedings of International Workshop on Natural and Artificial Immune Systems (NAIS) Vietri sul Mare, Salerno, Italy, June 9–10, 2005.

657. J. Timmis. On Diversity and Artificial Immune Systems: Incorporating a Diversity Operator into aiNet. Published in the proceedings of International Workshop on Natural and Artificial Immune Systems (NAIS) Vietri sul Mare, Salerno, Italy, June 9–10, 2005.

658. J. Timmis, Camilla Edmonds. A Comment on opt-AiNET: An Immune Network Algorithm for Optimization. Published in the proceedings of International Conference on Genetic and Evolutionary Computation (GECCO), Seattle, Washington USA, June 26–30, 2004.

659. J. Timmis, Camilla Edmonds and Johnny Kelsey. Assessing the Performance of Two Immune Inspired Algorithms and a Hybrid Genetic Algorithm for Function Optimization. Published in the proceedings of the Congress on Evolutionary Computation (CEC). Portland, Oregon USA, June 20–23, 2004.

660. J. Timmis, Knight, T., de Castro, L. N. and Hart, E. An Overview of Artificial Immune Systems 'Computation in Cells and Tissues: Perspectives and Tools of Thought'. Edited by Ray Paton, 2001.

661. J. Timmis, J. I., Knight, T., L. N. De Castro and E. Hart. An Overview of Artificial Immune Systems: An Emerging Technology, invited chapter for the book CYTO-COM, 2001.

662. J. Timmis. aiViS - artificial immune network visualization. Published in the proceedings of EuroGraphics U.K. Conference, pages 61–69, University College London. April 2001. Eurographics.

663. J. Timmis and M. Neal. Investigating the evolution and stability of a resource limited artificial immune system. Published in the proceedings of Genetic and Evolutionary Computation Conference GECCO, Las Vegas, Nevada, USA, July 8, 2000.

664. J. Timmis and T. Knight. Artificial Immune Systems: Using the Immune System as Inspiration for Data Mining. Published in Data Mining: A Heuristic Approach. H. A. Abbass, R. A. Sarker, and C. S. Newton (eds.)

665. J. Timmis, M. Neal and T. Knight. AINE: Machine Learning Inspired by the Immune System. Published in IEEE Transactions on Evolutionary Computation, June 2002.

666. J. Timmis, M. Neal and J. Hunt. An artificial immune system for data analysis. Biosystems, 55(1/3): 143–150, 2000.

667. J. Timmis. Visualizing artificial immune networks. Technical Report UWA-DCS-00-034, University of Wales and Aberystwyth, 2000.

668. J. Timmis and M. J. Neal. A Resource Limited Artificial Immune System for Data Analysis. Research and Development in Intelligent Systems XVII, pages 19–32, December 2000. Published in the proceedings of ES, Cambridge, U.K.

669. J. Timmis. On parameter adjustment of the immune inspired machine learning algorithm AINE. Technical Report 12-00, Computing Laboratory, University of Kent at Canterbury, Canterbury, Kent. CT2 7NF, November 2000.

670. J. Timmis and M. Neal. Investigating the evolution and stability of a resource limited artificial immune system. In A. S. Wu, editor, Special Workshop on Artificial Immune Systems, Genetic and Evolutionary Computation Conference (GECCO), Workshop Program, AAAI Press, pp. 40–41, Las Vegas, Nevada, U.S.A., July 2000.

671. J. Timmis, M. Neal, and J. Hunt. Data analysis with artificial immune systems and cluster analysis and kohonen networks: Some comparisons. Published in the proceedings of IEEE Int. Conf. Systems and Man and Cybernetics, pp. 922–927, Tokyo, Japan. 1999.

672. J. Timmis, M. Neal and J. Hunt. An Artificial Immune System for Data Analysis. Published in the proceedings of the International Workshop on Intelligent Processing in Cells and Tissues (IPCAT), 1999.

673. I. Tizzard. Immunology: An Introduction 2nd Edition. Pub. Saunders College Publishing. The Response of B-Cells to antigen. pp. 199–223. 1988.

674. N. Toma, S. Endo and K. Yamada. The Proposal and Evaluation of an Adaptive Memorizing Immune Algorithm with Two Memory Mechanisms. Journal of Japanese Society for Artificial Intelligence. Vol. 15, No. 6, pp. 1097–1106. 2000.

675. N. Toma, S. Endo and K. Yamada. Immune Algorithm with immune network and mhc for adaptive problem solving. Published in the proceedings of IEEE International Conference on Systems and Man and Cybernetics (SMC). Tokyo, Japan, IEEE. 1999.

676. Krzysztof Trojanowski and Martin Sasin - The Idiotypic Network with Binary Patterns Matching. Published in the proceedings of the 5th International Conference on Artificial Immune Systems (ICARIS), Portugal, 4–6 September, 2006.

677. K. Trojanowski and S. T. Wierzchon, Memory management in Artificial Immune System, Published in the proceedings of ICNNSC 2002.

678. Jamie Twycross, Uwe Aickelin. Conceptual: Biological Inspiration for Artificial Immune Systems. In the proceedings of 6th international conference on Artificial Immune systems, 26th–29th August, 2007 in Santos/SP, Brazil.

679. Jamie Twycross and Uwe Aickelin. Libtissue-Implementing Innate Immunity. Published in the proceedings of IEEE World Congress on Computational Intelligence (special session on recent development in artificial immune systems) in Congress on Evolutionary Computation, Vancouver, Canada, July 17–21, 2006.

680. J. Twycross and U. Aickelin. Towards a Conceptual Framework for Innate Immunity. Published in the proceedings of ICARIS, 4th International Conference on Artificial Immune Systems, LNCS, Springer-Verlag, Banff, Canada, 2005.

681. A. M. Tyrrell. Computer Know Thy Self! : A Biological Way to look at Fault Tolerance. In 2nd Euromicro/IEEE Workshop on Dependable Computing Systems, Milan, 1999.

V

682. F. Varela, A. Coutinho, B. Dupire and N. Vaz. Cognitive Networks: Immune and Neural and Otherwise. Theoretical Immunology: Part Two, SFI Studies in the science of Complexity, 2, pp. 359–371. 1988.

683. Patrícia A. Vargas, Leandro N. de Castro, Roberto Michelan and Fernando J. Von Zuben. Implementation of an Immuno-Genetic Network on a Real Khepera II Robot. Published in the proceedings of the IEEE Congress on Evolutionary Computation, Canberra, Australia, December 8th–12th 2003.

684. Patrícia A Vargas, Leandro N. de Castro, Fernando J. Von Zuben. Artificial Immune Systems as Complex Adaptive Systems. In 1st International Conference on Artificial Immune Systems (ICARIS), University of Kent at Canterbury, U.K., September 9th–11th, 2002.

685. F. Vargas, R. D. Fagundes, D. Barros Jr. A New On-Line Robust Approach to Design Noise Immune Speech Recognition Systems. Published in the proceedings of the Eighth IEEE International On-Line Testing Workshop (IOLTW). Isle of Bendor, France. July 08–10, 2002.

686. F. Vargas, R. D. Faugundes, D. Barros Jr. Summarizing a New Approach to Design Speech Recognition Systems: A Reliable Noise-Immune HW-SW Version 14th Symposium on Integrated Circuits and Systems Design. Pirenopolis, Brazil. September 10–15, 2001.

687. M. Velez, F. Nino and O. M. Alonso. A Game-Theoretic Approach to Artificial Immune Networks (Conceptual Paper). Published in the proceedings of the Third International Conference on Artificial Immune Systems (ICARIS), Catania, Italy. September 13–16, 2004.

688. F. T. Vertosick and R. H. Kelly. Immune Network Theory: a Role for Parallel Distributed Processing Immunology, Vol. 66, pp. 1–7, 1989.

689. Cutello Vincenzo, Giuseppe Nicosia, Pietro Oliveto and Mario Romeo. On the convergence of immune algorithms. To appear in the proceedings of the First IEEE Symposium on Foundations of Computational Intelligence (FOCI) 1–5 April 2007, Honolulu, Hawaii, USA.

690. F. Vistulo de Abreu, E.N.M. Nolte-'Hoen, C.R. Almeida and D.M. Davis - Cellular frustration: a new conceptual framework for understanding cell-mediated immune responses. Published in the proceedings of the 5th International Conference on Artificial Immune Systems (ICARIS), Portugal, 4–6 September, 2006.

W

691. Joanne H. Walker and Simon M. Garrett. Dynamic Function Optimisation: Comparing the Performance of Clonal Selection and Evolution Strategies. Published in the proceedings of Second International Conference on Artificial Immune Systems (ICARIS), September 1–3, 2003, Napier University, Edinburgh, U.K.

692. Y. Wantanabe, A. Ishiguro and Y. Uchikawa. Decentralized Behavior Arbitration Mechanism for Autonomous Mobile Robot Using Immune Network. Chapter 10 in the book entitled Artificial Immune Systems and Their Applications, Publisher: Springer-Verlag, Inc., pp. 187–207, January 1999.

693. Y. Wantanabe, A. Ishiguro, Y. Shirai and Y. Uchikawa. Emergent construction of a behavior arbitration mechanism based on the immune system. Advanced Robotics, 12(3), pp. 227–242, 1998.

694. C. Warrender, S. Forrest and L. Segel. Effective Feedback in the Immune System. Published in the proceedings of Genetic and Evolutionary Computation Conference (GECCO). (Workshop Program.) San Francisco. California. July 7, 2001.

695. C. Warrender, S. Forrest and B. Pearlmutter. Detecting intrusions using system calls: Alternative data models. 1999 IEEE Symposium on security and Privacy, 1999.

696. Andrew Watkins, Jon Timmis, and Lois Boggess. Artificial Immune Recognition System (AIRS): An Immune Inspired Supervised Machine Learning Algorithm. *Genetic Programming and Evolvable Machines*, 5(1), March 2004. Link [http://www.cs.kent.ac.uk/pubs/2004/1634/]

697. Watkins & J. Timmis. Artificial Immune Recognition System (AIRS): Revisions and Refinements. In 1st International Conference on Artificial Immune Systems (ICARIS), University of Kent at Canterbury, U.K., September 9th–11th, 2002.

698. A. B. Watkins and L. C. Boggess. A Resource Limited Artificial Immune Classifier. Published in the proceedings of the special sessions on artificial immune systems in Congress on Evolutionary Computation, IEEE World Congress on Computational Intelligence, Honolulu, Hawaii, May 2002.

699. A. Watkins and J. Timmis. Exploiting Parallelism Inherent in AIRS, an Artificial Immune Classifier (Conceptual Paper). Published in the proceedings of the Third International Conference on Artificial Immune Systems (ICARIS), Catania, Italy. September 13–16, 2004.

700. L. Wenjian, C. Xianbin and W. Xufa. An Immune Genetic Algorithm Based on Immune Regulation. Published in the proceedings of the special sessions on artificial immune systems in Congress on Evolutionary Computation, IEEE World Congress on Computational Intelligence, Honolulu, Hawaii, May 2002.

701. Jennifer A. White and Simon M. Garrett. Improved Pattern Recognition with Artificial Clonal Selection. Published in the proceedings of Second International Conference on Artificial Immune Systems (ICARIS), September 1–3, 2003, Napier University, Edinburgh, U.K.

702. S. R. White, M. Swimmer, E. J. Pring, W. C. Arnold, D. M. Chess and J. F. Morar. Anatomy of a commercial-grade immune system. IBM Thomas J. Watson Research Center, Yorktown Heights, New York, USA, 2000.

703. Amanda Whitbrook, Uwe Aickelin and Jonathan Garibaldi (2007): 'Idiotypic Immune Networks in Mobile Robot Control', IEEE Transactions on Systems, Man and Cybernetics, Part B, 37(6), 1581- 1598, doi: 10.1109/TSMCB.2007.907334, Abstract [http://www.cs.nott.ac.uk/%7Euxa/papers/07ieee_smcb_ais_robot.html], Paper [http://www.cs.nott.ac.uk/%7Euxa/papers/07ieee_smcb_ais_robot.pdf].

704. S. Wierzchon & Kuzelewska. Stable Clusters Formation in an Artificial Immune System. In 1st International Conference on Artificial Immune Systems (ICARIS), University of Kent at Canterbury, U.K., September 9–11, 2002.

705. S. Wierzchon. Generating optimal repertoire of antibody strings in an artificial immune system. In Intelligent Information Systems, M. A. Klopotek, M. Michale-wicz and S. T. Wierzchon, Eds. Heidelberg, Germany: Advances in Soft Computing. Series of Physica- Verlag. pp. 119–133, 2000.

706. S. T. Wierzchon. Deriving concise description of non-self patterns in an artificial immune system. In: L. C. Jain, J. Kacprzyk, (Eds.), New Learning Paradigm in Soft Computing. Physica-Verlag 2001, ISBN 3-7908-1436-9, 438–458, 2001.

707. S. T. Wierzchon. Multimodal optimization with artificial immune systems. In: M. A. Klopotek, M. Michalewicz, S. T. Wierzchon, Eds. Intelligent Information Systems 2001. Physica-Verlag 2001, 167–179, 2001.

708. S. T. Wierzchon and U. Kuzelewska. An artificial immune network as a tool for data analysis and clustering. In: J. Rybicki, A. Tylikowski, eds, "Simulation in Search and Development", Proc. 8th Worshop of Polish Simulation Society, Computer Center TASK, Gdansk Technical University 2001, 421–426, 2001.

709. S. T. Wierzchon. Discriminative power of the receptors activated by k-contiguous bits rule. (Invited paper) Journal of Computer Science and Technology. Special Issue on Research Computer Science, vol. 1, no. 3, pp. 1–13, 2000.

710. William Wilson, Phil Birkin, Uwe Aickelin. Conceptual: Motif Detection Inspired by Immune Memory. In the proceedings of 6th international conference on Artificial Immune systems, 26th–29th August, 2007 in Santos/SP, Brazil.

711. William O. Wilson, Phil Birkin and Uwe Aickelin - Price Trackers Inspired by Immune Memory. Published in the proceedings of the 5th International Conference on Artificial Immune Systems (ICARIS), Portugal, September 4–6, 2006.

712. W. O. Wilson and S. Garrett. Modelling Immune Memory for Prediction and Computation (Conceptual Paper). Published in the proceedings of the Third International Conference on Artificial Immune Systems (ICARIS), Catania, Italy. September 13–16, 2004.

713. Jui-Yu Wu and Yun-Kung Chung. Artificial Immune System for Solving Generalized Geometric Problems: Preliminary Results. Published in the proceedings of the Genetic and Evolutionary Computation Conference (GECCO), Washington, D.C., June 25–29 2005.

714. Z. Wu and Liang Y. Self-regulating Method for Model Library Based Artificial Immune Systems. Published in the proceedings ICARIS, 4th International Conference on Artificial Immune Systems, Banff, Canada, 2005.

715. Zejun Wu and Liang Y. Self-regulating Method for Model Library Based Artificial Immune Systems. Published in the proceedings of ICARIS, 4th International Conference on Artificial Immune Systems, Banff, Canada, 2005.

716. Zejun Wu, Hongbin Dong, Yiwen Liang, R. I. McKay. A Chromosome-based Evaluation Model for Computer Defense Immune Systems. Published in the proceedings of the IEEE Congress on Evolutionary Computation, pp. 1363–1369, Canberra, Australia, December 8–12, 2003.

X

717. S. Xanthakis, S. Karapoulios, R. Pajot and A. Rozz. Immune System and Fault Tolerant Computing. In J. M. Alliot, editor, Artificial Evolution, volume 1063 of Lecture Notes in Computer Science, pp. 181–197. Springer-Verlag, 1996.

718. Ren-Bin Xiao, Lei Wang, Yong Liu. A framework of AIS based pattern classification and matching for engineering creative design. Published in the proceedings of the International Conference on Machine Learning and Cybernetics, pp. 1554–1558, Vol. 3, November 4–5, 2002.

719. Le Xu, Mo-Yuen Chow, Jon Timmis, Leroy Taylor and Andrew Watkins. On the Investigation of Artificial Immune Systems on Imbalanced Data Classification for Power Distribution System Fault Cause Identification. Published in the proceedings of IEEE World Congress on Computational Intelligence (special session on recent development in artificial immune systems) in Congress on Evolutionary Computation, Vancouver, Canada, July 16–21, 2006.

Y

720. Yingbing Yu, J. Graham, Soft Computing for Masquerader Detection Based Upon Artificial Immunity Model. Published in the proceedings of 8th World Conference on Systemics, Cybernetics and Informatics (SCI) Orlando, FL, July 2004, Vol. 17, pp. 85–90, Publisher—International Institute of Informatics and Systemics.

721. Shengxiang Yang. A Comparative Study of Immune System Based Genetic Algorithms in Dynamic Environments. Published in the proceedings of the Genetic and Evolutionary Computation Conference (GECCO), Seattle, Washington, July 8–12, 2006.

722. Yingbing Yu, J. Graham, Computer Immunology and Neural Network Models for Masquerader Detection from User Command Sequences, 17th Intl. Conference on Computer Applications in Industry and Engineering (CAINE), Orlando, FL, Nov. 2004, pp. 1–6.

Z

723. Yanchao Zhang, Que Xirong, Wang Wendong, Cheng Shiduan. An immunity-based model for network intrusion detection. Published in the proceedings of International Conferences on Info-tech and Info-net, ICII 2001, Beijing. Vol. 5, pp. 24–29. October 29–November 1, 2001.
724. Jian Zhang, Hua-Can He, Min Zhao. Hybrid detector set: detectors with different affinity. Published in the proceedings of the 3rd international conference on Information security, Shanghai, China, 2004, Pages: 87–91.
725. X. Zhang, G. Dragffy, A. G. Pipe and Q. M. Zhu. Artificial Innate Immune System: an Instant Defence Layer of Embryonics (Conceptual Paper). Published in the proceedings of the Third International Conference on Artificial Immune Systems (ICARIS), Catania, Italy. September 13–16, 2004.
726. Zeming Zhang, Wenjian Luo and Xufa Wang. Immune Genetic Programming Based on Register-Stack Structure. In the proceedings of Congress on Evolutionary Computation (CEC) 2007, 25–28 September, Singapore.
727. Junzbong Zhao, Huang Houkuan. An evolving intrusion detection system based on natural immune system. Published in the proceedings IEEE Region 10 Conference on Computers, Communications, Control and Power Engineering (TENCON), Vol. 1, pp. 129–132. October 28–31, 2002.
728. Zhen-Qiang Qi, Guang-Da Hu, Zhao-Hua Yang, Fu-En Zhang. A novel control algorithm based on Immune Feedback Principle. Published in the proceedings of the International Conference on Machine Learning and Cybernetics, pp. 1089–1092, Vol. 2, November 4–5 2002.
729. Marcos Zuñiga, María-Cristina Riff, Elizabeth Montero. Search and Optimization: CD-NAIS: A Calibrated Immune System to Solve Constraint Satisfaction Problems. In the proceedings of 6th international conference on Artificial Immune systems, 26th–29th August, 2007 in Santos/SP, Brazil.
730. Xingquan Zuo. Robust scheduling method based on workflow simulation model and biological immune principle. Published in the Proceedings of the 9th annual conference on Genetic and evolutionary computation (GECCO) 2007, pp. 187–187, London, England.

Index

Milton Keynes UK
Ingram Content Group UK Ltd.
UKHW040446071024
449327UK00020B/1031